EVERY PERSON'S
LITTLE
BOOK OF
P=L=U=T=O=N=I=U=M

Stanley Berne with Arlene Zekowski

This non-fiction book is about "wild science and rogue technology." It exposes, to a popular audience, the "other side" (some call it the "dark side") of the "Nuclear Business."

The book addresses itself to every person in the U.S. who is now forced to assume an unwitting "stake" in both the weapons industry and in nuclear power. And the stakes are high. Each of us is being exposed, on a daily basis, to nuclear explosions, nuclear leaks, and we are drinking water and eating food laced with nuclear contamination.

The villain is P=L=U=T=O=N=I=U=M. It is a waste product which gathers at the heart of every nuclear reactor anywhere in the world, including the 112 nuclear plants in the U.S., and in the federally-owned nuclear weapons facilities around the country.

P=L=U=T=O=N=I=U=M is an indestructible *man-made element which governments and terrorists everywhere lust after, since it is the essential trigger in all Nuclear Bombs.*

For the first time, this book explains in clear, straight-forward, non-technical, popular language:

What P=L=U=T=O=N=I=U=M is. Who makes it? What it does. Where it is found. Who has it? Who buys and sells it? Why is it so deadly? How long does it last? Who needs it? Why is it called "Gray Gold"? Have you breathed it? Why are YOU paying billions for it . . .??????

every
person's
little
book
of
P=L=U=T=O=N=I=U=M

Books by Stanley Berne

THE GREAT AMERICAN EMPIRE

FUTURE LANGUAGE

THE UNCONSCIOUS VICTORIOUS AND OTHER STORIES

THE MULTIPLE MODERN GODS AND OTHER STORIES

THE DIALOGUES

THE NEW RUBAIYAT OF STANLEY BERNE

CARDINALS AND SAINTS

A FIRST BOOK OF THE NEO-NARRATIVE

EVERY PERSON'S LITTLE *BOOK OF P=L=U=T=O=N=I=U=M*

Anthology Inclusions

AMERICAN WRITING TODAY

BREAKTHROUGH FICTIONEERS

FIRST PERSON INTENSE

TRACE (1965 Series)

NEW WORLD WRITING (No. 11)

EVERY PERSON'S LITTLE BOOK OF P-L-U-T-O-N-I-U-M

Stanley Berne
with Arlene Zekowski

Drawings by Frank Salcido

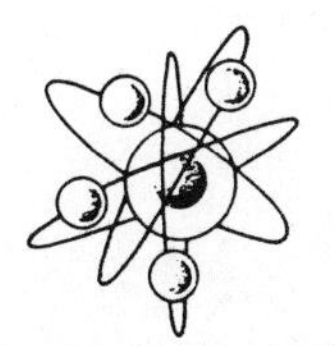

Rising Tide Press
P.O. Box 6136
Santa Fe, NM 87502-6136 USA

Library of Congress Catalog Card Number: 91-62794

ISBN: 0-913844-19-5

First Printing: 1992

Manufactured in the United States of America

The authors wish to express their thanks to the following
people who gave their generous help to this project:
Chris Wentz, Judi West, Eve Muir, Ann Sandefur,
Judy Backe, Daryl Cord, Becky Wilentz, Randall G. West.

Typography and Typeset : Carol Delattre
Illustrations and Cover Drawing: Frank Salcido

RISING TIDE PRESS
P. O. Box 6136
Santa Fe
New Mexico 87502-6136 USA

Through the release of atomic energy, our generation has brought into the world the most revolutionary force since the discovery of fire.

This basic power of the universe cannot be fitted into the outmoded concept of narrow nationalisms. For there is no secret and there is no defense; there is no possibility of control except through the aroused understanding and insistence of the peoples of the world.

—Albert Einstein

The human race has today the means for annihilating itself—either in a fit of complete lunacy, i.e., in a big war, by a brief fit of destruction, or by careless handling of atomic technology, through a slow process of poisoning and of deterioration in its genetic structure.

—Max Born

Some ten million times each year you ventilate about a pint of atmosphere into your lungs.

—Jon R. Luoma

TABLE OF CONTENTS

Table of Contents Continued

AUTHOR'S PREFACE

As a soldier in World War II, I left the Philippine Islands after we won it back from the Japanese, and took ship with General Douglas MacArthur, traveling to Japan, to become part of the American Army of Occupation. I was wounded in the Philippines, but being only nineteen, recovered quickly and stayed with my company, which was a part of what then was called The Signal Corps.

As part of my duties in Japan, I was assigned to visit Hiroshima. Of course, this was well after we bombed it and the Japanese surrendered.

Being a young man, and having already seen so much destruction in Manila and the Pacific Islands, I saw the devastation of Hiroshima as yet another absurdity of war. The images I remember, still, are the ones I describe in this book.

But to this day, I can still see the human devastation and pain caused by the Nuclear Bomb.

I can see again the long line of Japanese victims waiting patiently to enter the field hospital that we had set up to treat the burned and wounded.

What sickened me at the time was to see women, children and old men being dragged along in what appeared to be little four wheeled toy wagons that children play with. Those who could still walk, walked, and some of them dragged family members behind them on the little wagons. I saw the skin burned off their backs, and hanging down in peeled strips and ragged pieces, barely covering the red raw skin underneath, many with bones showing in their backs. They were all walking corpses. It was eerily silent. No outcry, no voice sound. They seemed to have no strength to speak, and they all appeared resigned to death. Of course, they soon did die.

It was only years later that I came to realize what had happened. A young soldier, even a beginning author as I was then, has very little powers of judgement or introspection. When you are nineteen, everything that happens to you in war is exactly like sleepwalking through a dream. All you can know, really, is that you are alive, somehow, and that you want to go home in order to be back in familiar surroundings. Above all, you want to get on with your life. War is an interruption to living. And being young, you want only to get back to life.

Today, after a long career as an author, I took an entirely new tack in my writing in order to write this book. The reason for writing this book is that my close proximity to Los Alamos, New Mexico (My home is in Santa Fe.) allowed me to see close-up, the sad direction we have taken *since* WW II. That war ended as the result of a few hand-made Nuclear Bombs put together at Los Alamos. What was the total nuclear arsenal in the world at that time? Perhaps less than four bombs.

Today, the Department of Energy directs the mass manufacture of Nuclear Bombs. It is now—Bombs off the assembly line, like cars.

Having seen first hand, the pain, the suffering and the blow-away destruction caused by just *one* bomb, I shudder to think what lies in wait for us in the future.

I came to see P=L=U=T=O=N=I=U=M as the personification, the very paradigm of poisonous pollution. I thought, if we could come to grasp the significance of this single deadly filth of our own making—if we could come to grips with it and understand it fully, then there may be hope that that knowledge might lead us to demand leaders who would exercise the power of this great country toward limiting and containing Nuclear Power in all its forms, all over the world.

Failing to do this, must surely condemn us and our children to the inevitable painful end that ignorance has in store for us.

Stanley Berne
Santa Fe, New Mexico
September, 1991

every
person's
little
book
of
P=L=U=T=O=N=I=U=M

ONE

AN ALIEN AND DANGEROUS WORLD TRANSFORMED

EARLY ON IN THE HISTORY of the world *everything* was radioactive. You and we would have found the world alien and dangerous. Radioactivity means that earth and rocks and fluids are alive and emitting invisible radio waves that have the power to force their way through our outer layers of skin and enter into our blood stream.

We are alive by virtue of our cells, which when first formed were charged with electricity enough to have each nucleus of each of our cells surrounded by orbiting moons that are in constant motion. To be alive means to be in constant internal orbit. Everything within us is whirling and turning. We are little earths surrounded by captive moons. Everything in the universe is made on the same pattern, and we are exact replicas of a whirling, orbiting, kinetic universe.

When you and we finally arrived here (courtesy of our grandmothers and grandfathers), the earth had finally had time to cool down. Radioactivity exists in

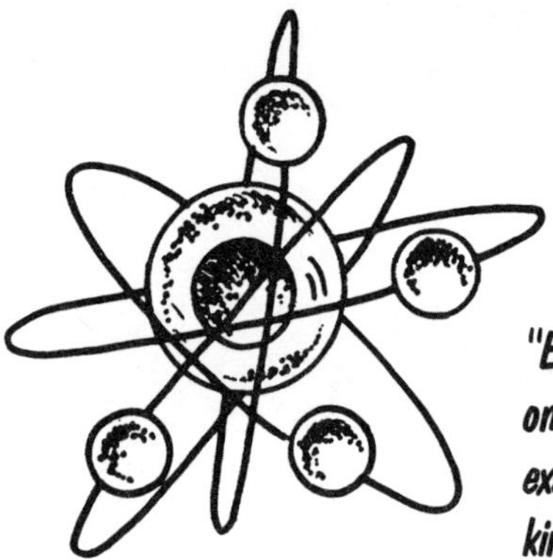

"Everything in the universe is made on the same pattern, and we are exact replicas of a whirling, orbiting, kinetic universe."

"half lives,"[1] which simply means that the universe, in constant flux and made up of radioactive elements, is slowly deteriorating and changing its form until its "half-lives" are finally over, and the given element

---

[1] A radioactive element decays at a regular rate. The measurable period of time it takes to decay is what we have come to call its "half life." In radium, for example, half of its atoms will have decayed into lead in 1,660 years; half of what is left, will again break down in another 1,660 years and so on until all its radiation is spent. There are short half-lives and long half-lives. The half-life of radioactive fluorine is eleven seconds, while the half-life of uranium is 4.5 billion years.

"dies." It no longer emits active radio rays. It has "deteriorated" into coal, lead, copper, gold, silver, iron, etc.

Not *all* radioactivity is finally over, however. We are still the creatures of elements formerly whirling through space. When the earth finally cooled, over eons of time, only then could life begin to exist, feed and reproduce. We are, each of us, still radioactive. When we pass a radiation monitor over our bodies, over our faces and necks, we hear and can measure faint signals of radioactivity. We are still radioactive, and we shall remain so, as a reminder of our origins of long ago.

TWO

THE POWER OF THE SUN

HUMANS ARE ANIMALS with great curiosity and an enormous appetite for discovery. This beautiful planet is our home and the center of our universe. While it is true that the whole earth was once radioactive, it cooled down so dramatically, that we have easily accommodated ourselves to what is now called "background radiation." The Earth cooled down, with the exception of certain deep pockets within the Earth where conditions were dry and isolated and stable enough to preserve radioactivity as a mirror of the former general condition of the Earth.

It is hard to say who "rediscovered" radioactivity.[2] In 1896 a French scientist, Antoine Henri

---

2 Radioactivity simply means "ray action."

Becquerel found that a certain element, Uranium, gave off both light and heat.

With that discovery, the great Nuclear race was on. Marie Curie and her husband Pierre found other elements which also retained their early charge of radioactivity. They discovered radium in 1910.

Many other scientists followed, measuring and naming elements of this old energy, newly discovered. Ernest Rutherford divided radioactivity into three different forms which he called Alpha, Beta and Gama.[3]

The question of radioactivity heated up at the end of the 19th century. It swelled into the most frenzied scientific quest of the modern era. Behind the rediscoveries, the namings, the measurements, lay the dream of a limitless source of energy and power for the world. Imagine harnessing unlimited power, a power related to the sun itself, a power of such magnitude and force that we would be enabled to transform our little Earth itself into heaven. There would never again be cold, or excessive heat, or hunger, or disease.

The people interested in this new subject were not only innovative investigators uncovering a thrilling new subject. They were people of the highest ideals—men and women who wanted to advance knowledge, to give the greatest gift science could bestow: a transforming power equal to the harnessing of the basic life force in the universe—the power within the central constructive element of the universe: the Atom.

---

[3] See opening of chapter 6 for a discussion of Alpha, Beta and Gamma.

THREE

A SEDUCTIVE AND SAVAGE BEAUTY

THE NUCLEAR AGE BEGAN on August 6, 1945 when the first atomic bomb was dropped on Hiroshima. 90,000 men, women and children were disintegrated outright, and another 90,000 were injured and died later, after a great deal of writhing, painful suffering.

Nuclear energy was clearly seen to possess ultimate military power. With the nuclear bomb, the whole world could be controlled. It was even predicted that the nuclear bomb would make war obsolete, since two adversaries, both armed with Nuclear, would cancel each other out, the destruction being too great a cost to bear.

The United States also became interested in the

"peaceful" uses for nuclear energy. A nuclear power plant is no different from any other electrical generating plant. To produce electricity, you need to boil water, produce steam, and then have the steam turn the propeller blades of a turbine that turns a shaft which produces an electrical current. You then channel the current over wires to your customers, and charge accordingly. Selling electricity to us, in the modern world, is to sell an inevitable product to a captive audience, thus producing uninterrupted wealth for the Utility Corporations.

Instead of burning coal or oil to produce steam, you use Uranium. You don't burn it, you "fission" it. Half inch long pellets of Uranium are packed into tubes made of zirconium, and 50,000 zircalloy fuel rods make up a reactor core. Control rods of boron are pushed in among the tubes to keep the neutrons calm and quiet. When you slowly pull the boron rods out of the way, the neutrons go wild—a chain reaction has begun.

Back in the 1930's Enrico Fermi of Italy broke apart the nucleus of the atom and found that that action produced energy. Leo Szilard, in London, split the atom and produced the first chain reaction. Working together in 1939, Szilard and Fermi produced the first practical chain reaction, a reaction that produced usable energy. Then they both died of radiation induced cancer.

There is a seductive and savage beauty to atomic energy. The energy of radioactivity, the first and overwhelming energy of our early universe, lies embedded still, in Uranium, that ancient substance harking back to the beginnings of our little Earth. It is as difficult to uncover and extract, as it is to uncover and extract gold. Both substances lie in deep pockets of the Earth. They have to be mined, extracted from deep veins, and they both lie mixed with ordinary earth and rock. They have to be processed, in order to purify them for commercial and industrial use.

One pound of Uranium-235 (U-235), the essential fuel in nuclear plants, is very scarce, only 1% of all raw Uranium ore. But a piece the size of a golfball can produce as much energy as 1500 tons of coal.

That's what makes Nuclear so seductive.

FOUR

THE DIVINE MYTH OF LIMITLESS ENERGY

THE THEORY OF UNIVERSAL unlimited power is more than seductive. As fossil fuels (coal, oil, natural gas) are used up and are irreplaceable, we ask ourselves how we shall produce enough steam to heat the water to turn the turbines so that we may continue to light the lights.

We are hopelessly dependent on power, for our machines that feed on power, for power coming at us over the wires from central power plants. We do not accept the limits of nature—cold, hunger, sickness. The life force within us demands food. We must transport ourselves and our possessions from one place to another. We must be warm in our homes. We must have food trucked to the markets where we can buy meats, apples, oranges, juices.

But the arcane secret society of nuclear engineers keeps its secrets well hidden from us. The influence of large utility corporations, whose product is power, overwhelms the government, and certainly overwhelms the population, and plans on pursuing the divine myth of limitless energy. Limitless energy, seductive and sweet as a dream, has behind it two savage costs: Radiation, and P=L=U=T=O=N=I=U=M-infected waste.

We are told nothing about the production of the filthiest substance known to man: P=L=U=T=O=N-=I=U=M. Every nuclear reactor works in a temperature that exceeds 2000º F. It therefore requires being bathed in water. That is the reason that Nuclear Power Plants are built on the banks of rivers and lakes. Cool fresh water is sloshed around the reactor in order to keep back fiery meltdowns. But the water becomes filthy with radiation,

"That is the reason that Nuclear Power Plants are built on the banks of rivers and lakes."

and then may be sloshed back into the river or lake from which it came. It settles P=L=U=T=O=N=I=U=M into the flesh of fish. They die. It sinks to the bottom and enters into the nearby water table. We take our drinking water from that table.

Nuclear reactor water is also cooled in huge cooling towers that give Nuclear Plants their characteristic profile, but 1% of that boiling contaminated water evaporates P=L=U=T=O=N=I=U=M into the air, so that we have to share our air with Nuclear.

Nuclear Power is savage and beautiful. It harnesses the deepest forces of nature, the power within the atom. It argues with us for its use: 6300 tons of Uranium is equal to 425 million barrels of oil—6300 tons of Uranium is equal to 120 million tons of coal—6300 tons of Uranium is equal to 2.6 trillion cubic feet of natural gas.

Oil, coal and natural gas are being used up and one day will be no more.

FIVE

MOTORS, SCRUBBERS, BLOWERS AND SUCKERS

RALPH WALDO EMERSON one of America's visionary early philosophers and poets, said of himself that he was really no better than anyone else, when it came to being "good." The difference was, he said, he was never sorely tempted. Therefore, he was able to remain "moral."

Our government is surely made up of good men and women, with children and families, and with as much stake in the future of America as we all have. We all want a country that is clean, where the air is breathable, the water drinkable. We want a beautiful country without suffering or painful burning cancers, caused by poisons.

But, we are sorely tempted to do evil. The terrible savage beauty of Nuclear Energy promises that, when fossil fuels are gone, we shall all enjoy the unlimited energy derived from the atom.

It is like making a pact with the Devil. The Devil smiles, is genuinely polite, and likable. He offers us wealth and prosperity, if only we will trade him our health, for unlimited riches. With Nuclear, the water will always boil; the steam will propel the generators; the electricity will forever flow: thus, we need not be at the mercy of foreign peoples who still have barrels of oil buried beneath their sands. We can be energy rich, yet free and independent.

The Devil is a formidable organizer and debater. He hires scientists and engineers of great skill, who are very methodical workers and planners. He is, however, a rather poor money manager. While we all need to plan and budget our money very carefully, limiting our spending, going without the new watch, or spending another winter in our worn coat, he secretively gathers together scarce resources, and builds great systems of buildings for the incineration of Nuclear Waste.

His managers ceremoniously piloted us through such an ingenious system. It was built in a small and efficient city of his own design. There, everything is planned for weapons, for contained explosions, for nuclear triggers, for the production of P=L=U=T=O=N=I=U=M, which is the chief ingredient of the atom bomb.

Uranium is turned into P=L=U=T=O=N=I=U=M. Every Nuclear Reactor burning Uranium, produces it as a waste product. When, during a chain reaction, a neutron happens to strike a nucleus of Uranium-238, the substance produced by such a strike is P=L=U=T=O=N-I=U=M-239. As the Nuclear Reactor continues to generate heat and produce steam, P=L=U=T=O=N=I=U=M-239 begins to accumulate at the core.

The manager-in-chief of the Nuclear Incinerator was a handsome, young, blond engineer, tall and athletic-looking, well-educated and clean cut. A real red-blooded young American. He was instantly likable. You would want him as a son, or a good friend. He let slip, however, during only one unguarded moment in our conversation, that his goal was to make as much money as possible, and he had carefully evaluated all the options that lay open to him as a young engineer, to turn his skills into as large a salary as he could command. The opportunity came when the government, interested, in this case, in continuing the production of atom bombs, offered him a position as chief engineer of their Nuclear Incinerator.

A Nuclear Incinerator is a formidable four story building full of stainless steel plumbing and tanks. It is full of motors, scrubbers, blowers, suckers, wells, tanks of coolant liquids, and is filled with ingenious elbow turns, funnels and pipes that wend their inevitable way toward a giant smoke stack, 10 stories high in the sky. All along the tanks, tubes and stacks, on the walls, in the floor, around the grates, behind the buckets, staring out at us everywhere, are the faces and dials of meters capable of reading out the measure of deadly radioactivity.

The Devil well knows that the deadliest product, or energy source thus far revealed to man, is P=L=U-

=T=O=N=I=U=M, which is hopelessly and poisonously radioactive. We are talking here about the creation of vast quantities of deadly radioactive waste, generated by the chemical and physical processes of producing P=L=U=T=O=N=I=U=M from Uranium products. P=L=U=T=O=N=I=U=M is the most highly radioactive substance on the face of the Earth. One breath of one granule of P=L=U=T=O=N=I=U=M will bring death by the slow painful inner melting of soft tissue. P=L=U-T=O=N=I=U=M must always be "remote-handled." That means that human arms must be shielded in lead-lined canvas sleeves and gloves, and as the deadly material is being shaped and handled from behind heavy triple glass windows, all of it is encased in "glove boxes" that are isolated from direct human contact with flesh. The devil of the substance that is P=L=U=T=O=N=I=U=M, must be operated upon without direct human skin or lung intervention. There are substances, however, that eat away at the seals, pipes, tubes and devices used in remote-handling. So, there is the constant danger of radiation escaping into the atmosphere, through burned-out seals and fittings.

Waste generated by the processing of P=L=U-=T=O=N=I=U=M is often mundane and unbelievably common. It is gloves, wrenches, hammers, screw-drivers, shoes, socks, coveralls, hats, masks, goggles, greasy wipes, rags, underwear—the contaminated (and actively radioactive) drapery of workers who are apparently paid enough to encourage them to come near all this radioactivity, without, hopefully, sucking up any of it. All the members of the working team, the supervisors as well, wear read-out radiation badges. These badges are read, from time to time, to determine the dose of radiation received by the host. When the limiting dose has been absorbed, the worker must be replaced by another worker who will be prepared to tolerate the next dose.

The devil of radiation is that you, or we, or the worker in the plant, cannot detect it. It is odorless, tasteless, invisible. It operates like radio waves, or television signals. You or we will never know when we are being irradiated. There is nothing about it that is detectable to the senses.

What is most unpleasant about P=L=U=T=O=N-=I=U=M is the menu of diseases associated with it. Here is a partial list of known effects:

—Bone marrow disease
—Blood production disturbances
—Heart attacks
—Lung diseases
—Pain in the bones and joints
—Chronic fatigue
—Skin lesions
—Hodgkin's disease
—Myelofibrosis
—Leukemia
—Absence of blood coagulant
—Acute abdominal attacks
—Stomach tumors
—Lung hemorrhaging
—Lumpy growths on the skull
—Large and small tumors (some the size of a hen's egg)
—Pain and weakness
—Chronic headache
—Breathing difficulties

—Severe pain in the heart, legs and chest
—Shortness of breath

P=L=U=T=O=N=I=U=M is attracted to human lungs and other internal soft tissue. P=L=U=T=O=N= I=U=M decays so slowly that it takes 24,000 years for half of its alpha radiation to decay. In other words, once produced, it cannot ever be disposed of, so the world's accumulation is simply added to and added to.

Other known effects:

—Chronic anemia
—Enlarged testes
—Disintegrating testes
—Decades of chronic lung disease
—Intestinal "attacks"
—Chronic weeping skin sores and ulcerations
—Polycythemia Vera (an excess of red blood cells)
—Tumors of the hip
—Skin cancers of the ear
—Testicles have to be removed
—Lung cancer that attaches itself to the heart
—Inability to breathe without applied oxygen
—Chronic respiratory illnesses
—Factor VIII (lack of blood coagulant)
—Macroglobulinemia (cancer of the bone marrow—an extremely rare form of cancer)
—Multiple myeloma (a brutally painful bone marrow cancer)

SIX

THE EFFECT OF RADIATION ON LIVING THINGS

AT THE BEGINNING of the 20th century, the physicist Ernest Rutherford identified three forms of radiation which he named after the first three letters of the Greek alphabet, *alpha, beta* and *gamma*.

Alpha rays, or "particles," are the least penetrating of the three charges. They are easily blocked by unbroken human skin, even by a sheet of paper.

Beta particles are smaller and lighter than alpha particles but move faster and with more energy. They have more penetrating power and can travel deeply into the human body.

Gamma rays are the most powerful of the three

invisible rays. They can be stopped, but it takes a lead shield at least half an inch thick to do so. They easily penetrate skin and tissue and can end up depositing in human bone and organs. This penetrating radiation, when it strikes human cells, can permanently alter the cells' electrical charge. This is called Ionization.

We were invited to attend a seminar on the medical management of radiation. It was given by a team from the Oak Ridge Associated Universities, and was sponsored by the Department of Energy (DOE).

In the course of the seminar, we were shown the case of the unfortunate "Juan L," from Juarez, Mexico, a terribly poor man who scraped a meagre living out of turning over the contents of the city dump, in his daily search for salvageable items.

On a certain day, he found several very beautiful and shiny rods tightly woven about with what appeared to be stainless steel thread. The rods were about six inches long and Juan put them into his trouser's pocket for safe keeping. He apparently kept the flexible wires there for several days, shifting them about from his right side pocket, to his left.

What Juan had no way of knowing was that the wires he had in his pockets were lethal parts of a discarded X-ray device that somehow ended up in the city dump.

The lecturer from Oak Ridge, an expert on radiation, showed us slides of Juan taken after he was admitted to the hospital. Juan was a young man of about 24, short, fat and swarthy, with deep radiation burns on both outer thighs. The lesions were large and weeping, with a variety of circular colorings from light pink to darkest red. The lesions were about a foot long each, and eight inches wide. Apparently, the bone structure on each leg had been well penetrated, and his penis and gonads were terribly swollen, and blood red in color. His sex organs had swelled to three times their normal size, and would eventually be amputated.

When we saw Juan, in a later slide, both legs had been amputated, and the skin from the back of his legs was very skillfully sewn forward over the area of the leg stumps, as if the skin was a diaper. It was neatly stitched into place.

The lecturer failed to tell us if Juan was now dead or alive, but from the look of him, it appeared that death would be a God-given mercy for this poor and innocent victim of radiation.

X-rays dose us with radiation. They are, for that reason, very carefully timed, and the technician must be carefully lead-shielded. A German physicist, William R. Roentgen, discovered X-rays in 1895. The measure of radiation, the "Roentgen," describes the effect of radiation on living things, but it is generally called REM, for "Roentgen Equivalent in Man." The average American absorbs 0.125 Rems a year from medical and dental X-rays. We become something of a permanent repository of Rems, once we have been dosed, and they remain with us until we die. There is no way they can be expelled from the body. We absorb 0.125 Rems a year from Background Radiation.

In 1960, the U.S. Environmental Protection Agency set a limit of absorption of 3 Rems per person, during any 3 month period, as the maximum dose the human body should absorb. That comes to 12 Rems a year. The Nuclear Regulatory Commission (NRC) set its yearly dose for nuclear plant workers at only 5 Rems a year.

"We become something of a permanent repository of REMS, once we have been dosed, and they remain with us until we die."

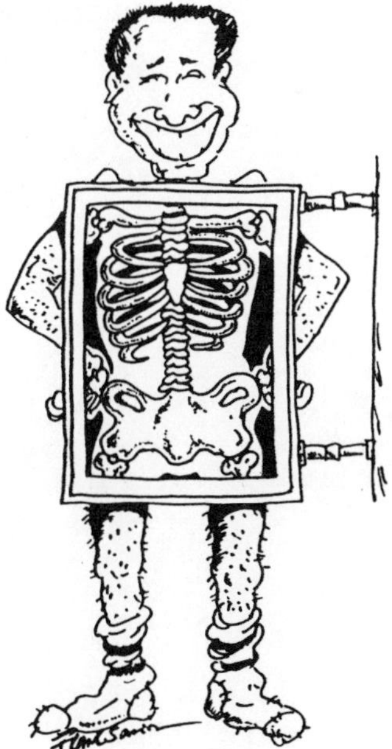

—A dose of 3000 Rems means instant death
—1000 Rems kills in 30 days
—450 Rems causes death in approximately half of such exposed victims
—The other half, at 450 Rems, either will become ill, and recover, or may never be well again

—250 Rems causes slow-moving cancers, and results in eventual death
—Below 250 Rems, there may result leukemia, tumorous cancers, and permanent seed damage to female eggs or male sperm, that show up in children demonstrating mental or physical handicaps
—There is no presently known, absolutely safe radiation level.

SEVEN

THE FINAL SOLUTION

SOME OF OUR MOST GIFTED and best trained young Americans are working in the defense and nuclear industries. They are outstanding, inventive, highly motivated, well paid, and quite willing to view the world from the point of view of their employers—the very large multinational corporations that are the weapons designers, weapons manufacturers, plant designers and exporters, incinerator waste fabricators and managers, for the nuclear industry.

They like to call what they do "science," but it is really only "technology."[4]

[4] "Science," in the classical sense, has to do with grand purposes and lofty ideals. Like Philosophy or Art, it partakes of the highest pursuits of mankind, which includes the state or fact of *knowing*. Science is always in pursuit of *knowledge*.

At Los Alamos, New Mexico, they have designed and built a P=L=U=T=O=N=I=U=M incinerator, with a price tag (for the next one, to be sold to the states, and to be exported abroad) that may well exceed $30 million, that is supposed to do away with the legitimate fear people have to burying nuclear waste. There have been many attempts made to bring the nuclear waste problem to a "final solution." A "final solution" has yet to be announced.

Their latest "final solution" is to burn radioactive waste. If they can't store it underground, where it melts and gases itself into future explosions, then they have decided to try to burn P=L=U=T=O=N=I=U=M, thus effectively burying it in the sky.[5]

"Technology" is the study of the practical, or the industrial. It deals with limited areas, such as the handling of specific technical problems. Making a toaster or a bomb are not "scientific" enterprises, although they may borrow applications of science such as the systematizing of facts.

"Engineering" applies scientific discoveries to practical use, such as the planning, design, construction and management of machinery and related systems.

---

[5] The Department of Energy (DOE), and their predecessors, have been seeking the "final solution" for over 40 years. The DOE has just announced a *new* final solution, although even they have their doubts. It's called "vitrification," which they have borrowed from the French. They are now going to encase radioactive waste, both liquid and solid, in glass logs wrapped in steel. To that end, they have just completed and dedicated a new $1.3 billion plant (called a Defense Waste Processing Facility) at Aiken, South Carolina. (Another new factory is under construction at West Valley, New York.) The Aiken plant was dedicated November 28, 1990.

W. Henson Moore, Deputy Secretary of Energy admitted that nuclear wastes in barrels may produce explosive gases and chemicals that can be spread by explosion or earthquake.

According to the DOE, glass logs cannot explode, but it is not yet certain how they will act when they crack or otherwise deteriorate.

Other "final solutions" have also been announced. One, is to make radioactive cement out of liquids, and shape them into blocks; the other, is to "evaporate" (again, read incinerate) wastes to a dry residue powder.

The difficulty with this solution, is that 1% of P=L=U=T=O=N=I=U=M is added to the air we breathe, and 1% a day, 24 hours a day, along with the grand plan of allowing the Giant Corporation to sell nuclear incinerators to every state in the union, and to every foreign country in the world, will very soon make our air hostile to life. Selling incinerators is at least as profitable as selling nuclear plants themselves, which they also build and export. The air of the world will be mixed with spirits of P=L=U=T=O=N=I=U=M. And that is going to prove fatal to our planet.

Everything about a Nuclear Power Plant is radioactive. *Everything!* The center of power in the plant is the reactor core. The core contains Uranium, P=L=U-=T=O=N=I=U=M, Krypton-85, Strontium-90, Iodine-131, (which is especially attracted to human thyroids and goes right after them),Cesium-137, Iron-59. Everything in contact with the core is Radioactive: the water to cool the reactor, the steam the reactor produces to move the turbines, the myriad of pipes and steel vessels in which water is heated and stored—everything is radioactive. The whole nuclear machine is surrounded by a 9-foot thick container of reinforced concrete to contain gamma rays. The containing structure has a domed outer wall of thick reinforced concrete.

The guts of the plant are the innumerable elaborate and intricate seals, valves and steel welds that can number in the millions, and that can breach at any time

---

We must keep in mind that none of the above processes does anything to reduce radioactivity. They just change the "shape" of the waste. Billions of dollars are being spent, and we end up with exactly what we had to begin with: *indestructible P=L=U=T=O=N=I=U=M.*

to allow unplanned emissions: a gasket can blow out; defective or worn seals can no longer contain radioactive water; leaking pipes drip radioactivity in the form of water, steam or lubricants; faulty wiring can prevent an accurate readout of internal conditions inside the reactor, where no man (or woman) may enter for fear of instant death.

Missing or broken parts can stop the production of an entire plant, or produce a major disaster—this latter is the story of Three-Mile Island.

There is also plain old human error. In October, 1978, the plant manager at the Federal Government Laboratory in Idaho, was watching the World Series on TV, instead of reading the tell-tale signs of a major impending disaster. His instruments were warning of a startling rise in temperature at the nuclear core. Instead, he was watching the Yankees drive in a home run. He threw the switch to stop the chain reaction, and fled the control room.

A recent advertisement by the U.S. Council for Energy Awareness, a front organization for the Nuclear Industry, warns us that 112 operating Nuclear Plants are not enough. There are now 112 Nuclear Reactors producing P=L=U=T=O=N=I=U=M in the U.S. There are well over 250 Nuclear Power Plants throughout the world. All of them have smokestacks that reach out into the sky.

Nuclear Power Plants produce P=L=U=T=O=N-=I=U=M, but they also produce Krypton-85. Released as a gas, K-85 released accidentally in 1991, won't become harmless until the year 2191. Thus, Krypton will be active in the air we breathe for the rest of our lives, as well as during the lives of our children, and for several generations of their children as well.

The half-life of Strontium-90 is 28 years; Carbon-14 will live for 5,770 years; Nickel-59 will live for 80,000 years; these devastatingly immortal effluences do not ever go away. They love to be near people. They don't accept the reality that they aren't welcome. They are dumb enough (or smart enough) to just "hang around."

Iodine-131, for example, loves ocean water, where it seeks out seaweed. The shrimp we all love to eat, browse on the seaweed. Thus, Iodine-131 enters directly into their little bodies. With fish, it is the same. Some fish can only survive in and among the forests of seaweed. You and we eat the shrimp and the fish; thus, Nuclear Iodine, from some nuclear power plant far away, ends up on our dinner plate.

Iodine-131 is a very active and volatile element. It loves to migrate. You can't stop it or kill it. It has the vitality of the Devil himself. It loves the human thyroid gland. It rides about in our blood stream until it arrives safely at its ultimate destination, the human thyroid gland, and it stays there forever, an unwelcome guest feeding on our vitality—for as long as we have any.

Strontium-90 loves milk. It gets to the udders of cows through the grasses upon which the cow has browsed. Children drink the milk and Strontium-90 immediately migrates to their teeth and bones. One suspects that a nursing mother, downwind of a Nuclear Plume will find an unwelcome guest at her breast, beside her own little one.

Radioactive Iron-59 migrates to our red blood cells and sails on through the circulatory system, distributing its poison to each cell.

Cesium-137 loves muscle tissue. It behaves the same way as does Potassium, only it contributes nothing to the body as Potassium does. Instead, it breaks down muscle fiber until we can no longer stand on our own two feet.

Krypton Gas is the most secretive of all elements. It is the ultimate "spy gas." It rides aloft in the air streams, high above the Earth, and thus gets to travel great distances. We don't know its ultimate goal, or where it is most likely to settle. We just don't know the mission of Krypton, and where or how it will end up—and what it will do to us when it gets there.

EIGHT

IS THAT ANY WAY TO RUN A BUSINESS?

EVERYTHING ABOUT A NUCLEAR POWER Plant is "out of this world." It is a dream world feeding on itself. It has an internal environment that is as alien to humans as space is. The space program operates on a percentage basis. So many flights may produce so many accidents. It is their cost of doing business. They know the dangers, and the people involved are ready to live with life and death situations.

But Nuclear Power Plants go far beyond the involvement of a limited few brave astronauts who are ready to trade the fantastic opportunity of a ride into the outer reaches of the universe—with death.

Whole populations, cities, towns, millions of men,

women and children, billions and billions of dollars in property, are involved in whether or not a plant of this magnitude of danger will survive, or explode around them.

If the plant manages to survive, how do populations deal with the by-products of fission, the waste materials, the hot radioactive water that must constantly wash in and out of the plant, in order to keep its temperature below 2000° F? What about emissions, accidental or otherwise? Nuclear Plants must ventilate. They need air to operate in, just as we do. Radioactive steam is always seeking to escape, and it enters the atmosphere loaded with immortal indestructible poisons.

Then, there are the inevitable barrels of "transuranic waste": cartons, cans, gloves, overalls, tools, hats, broken valves encrusted with P=L=U=T=O=N=I=U=M, broken and discarded pipe, burned out meters, pipe elbows, grates, pieces of concrete, lubricants, rags, galoshes, shoes—everything a Nuclear Plant touches becomes radioactive forever and therefore capable of poisoning life.

Inside the plant, a spectacular ballet is in progress. It involves teams of workers, more like dancers. They can only work inside the plant near areas where fission is taking place for two minutes at a time. In one Illinois plant it took 350 men, each working *only a few seconds* , to repair a minor pipe elbow that might normally have been repaired in a few hours by two maintenance men. In a New York plant, a pipe broke that had to be re-welded, one of literally thousands and thousands of welds in any given Nuclear Plant. *It took 1700 welders a full six months to repair a single broken pipe.*

The problem is P=L=U=T=O=N=I=U=M and all the related radioactive gases and waters that a Nuclear Plant must be bathed in and that constitutes its internal environment. Everything in the plant is lethal to humans and lethal to life. Is that any way to run a business? Can we pay the price of constant and imminent danger in order to keep the lights on? There are other ways to light the lights, that we shall discuss later.

The Nuclear Power Industry employs "Jumpers." When a Nuclear Power Plant needs even routine maintenance, it calls for no less than human sacrifice. "Jumpers" are the poor who are willing to do a few minutes work in an alien atmosphere, for pay. They "jump" in, dressed in coveralls and gloves, with a wrench or screwdriver, turn a nut, or tighten down a bolt, then they "jump" out. When they are "out," everything on them must be removed and placed in a barrel, clothing and tools alike, for disposal outside the plant because everything on them is soaked in P=L=U=T=O=N=I=U=M. After undressing, the "Jumper" goes to the showers, washes his hair and body, gets dressed, gets paid, and leaves. But "Jumpers" leave with more than their pay. They leave carrying off a "dose" of radioactivity that has entered their blood stream. It may take as long as ten years before that "dose" surfaces in a cancer. Meanwhile, they have put bread on the table, so their children may eat. The Nuclear Industry considers such people disposable.

The use of "Jumpers" has been on the increase. For example, their number doubled between 1973 and 1976. The industry will publish no figures or details or studies on their use of "Jumpers." But, especially as plants get older and require more maintenance, more and more "Jumpers" will have to be used to do even simple routine jobs.

Nuclear Power Plants are constructed as carefully as possible. Every mile of pipe has been designed with both efficiency and safety in mind. All our American ingenuity and know-how has been sacrificed to these plants. The best young minds in America have been employed, the most intelligent of our young engineers are hired to see to it that the plants operate at maximum efficiency. If our American infrastructure has been deteriorating rapidly, it is because our tax money has gone to help build these plants, and our most creative young minds have been drained into this industry. Why? Because every Nuclear Power Plant produces P=L=U=T=O=N=I=U=M as a by-product of fission. And P=L=U=T=O=N=I=U=M is the trigger that makes the Atom Bombs explode.

Nuclear Power Plant accidents have, so far, been largely the result of human operator error. The Massachusetts Public Interest Research Group (MPIRG) has monitored and recorded power plant accidents. They discovered the following:

—A door was propped open for a week at a Nebraska plant and left that way, exposing workers to 5 or 6 Rems an hour.

—In Oregon, two technicians entered an unmonitored area. One man took a 17 Rem dose; the other took 27 Rems.

—In Wisconsin, a supervisor spent 30 seconds in a 2000 Rems an hour radiation-filled room, and never knew it.

—The Union of Concerned Scientists (UCS) reported a plant where the drinking fountain was accidentally hooked up to a contaminated water source.

—In another plant, a basketball was used to seal up a reactor pipe. When the basketball worked loose, 14,000 gallons of radioactive water gushed out behind the plant into the parking lot.

"Nuclear Power Plant accidents have, so far, been largely the result of human operating error."

Allow us to repeat that Nuclear Radiation is the most secretive of events. It never gives any warning of its presence. It is invisible, has no sound, we cannot taste it or feel it, and it is entirely odorless. There are well over 50,000 Nuclear Workers in the industry, and untold millions and millions of us near plants, research facilities, manufacturing centers, and poisonous dumps—all dealing out lethal radiation. A person receiving a fatal dose will never know what hit him or her. The insulting and painful results come later.

NINE

THE FRONT END OF THE NUCLEAR FUEL CYCLE

EVERYTHING ABOUT NUCLEAR POWER, used to boil water to produce steam that drives turbines that sends out electricity—everything about it is dirty, dangerous, poisonous and volatile. The essential fuel ingredient at the center of the process, is hot and radioactive Uranium: fissionable gold.

Uranium—secreted in old pockets of our early Earth's radioactivity—is to be found in South Africa, Australia, Russia, Canada and the U.S. In the U.S. the mines are found in the Western States of New Mexico, Wyoming, Colorado, Utah and Washington.

Uranium ore contains Radium, and is naturally radioactive, constantly emitting alpha particles, along with deadly Radon. In New Mexico, a public health medical doctor found an "epidemic of lung cancer among former miners." Many miners are American Indians, who have a hard time finding other work, so they drift into the fatal mines, many of which are located on Indian lands.

After mining, comes milling. In this activity Uranium is separated from its enveloping earth and rock to produce what is called "Yellowcake." To produce Yellowcake, Uranium is crushed in a large press and then washed out with water and chemicals, in an indoor factory that is unbelievably dirty and dusty and sopping with radioactive rinses and chemicals. Workers in the plant (many of them women), face exactly the same hazards as the miners.

Enormous waste piles or "tailings" are produced both at the mines and at the mills. Usually, for the sake of convenience, mills are built right next to the mines. There are now more than 21 Uranium mills in the U.S., and the quantity of tailings is enormous: we are getting close to 750 million tons of tailings a year, according to estimates by the Nuclear Regulatory Commission.

Tailings are just piles of radioactive waste thrown up behind the mills, where the wind can blow the tops off the piles, and the rain and snow can easily wash the radioactive granules off into the streams, rivers and underground drinking water systems, called aquifers.

Because of the danger of these growing piles of tailings, Congress enacted the "Uranium Mill Tailings Radiation Control Act of 1978." But it has seemed to have no effect whatever on the problem. The problem is enormous, and growing out of control. There was no provision made to enforce the law, or to have inspectors monitor conditions at the mills.

In 1979 (a year after Congress passed the law), a mill in New Mexico sent 1100 tons of tailings and 100 million gallons of radioactive water and chemicals into a small river known as the Rio Puerco. As a consequence, local wells and drinking water for miles around were poisoned. Many American Indian families live in the area, and it has been reported that a very high incidence of new-born babies have been diagnosed as mentally retarded. The plant was touted as a "model" factory, and judged by the Government to be "extra safe," and a creditable addition to the area, providing much needed employment for the Indians.

Once Yellowcake has been milled, it is shipped to Illinois and Oklahoma to be converted into a gas. These states have the only two gas-conversion plants in the U.S. Both plants are operated at taxpayer expense, by the U.S. Government. Next, the gas has to be trans-shipped again, this time to Tennessee, Kentucky and Ohio, to Federally operated and Federally owned plants, where it is "Enriched." Enriching means that the Uranium gas is treated with a variety of chemicals to increase its concentration of fissionable Uranium-235, which is then filtered down to produce another gas, which is now 97% Uranium-238, and 3% Uranium-235. In this form, it can now be employed to sustain a chain reaction at a Nuclear Power Plant. Simply put, it can now burn hot enough to boil water to produce steam.

"Enriching" produces still more radioactive

waste and tailings—tons and tons of it. For every 5 pounds of Yellowcake, only 1 pound emerges as usable Uranium-235. The other four pounds are left over as useless and dangerous radioactive poison.

The next process is "Fuel Fabrication." Uranium gas must now be re-converted into solid pellets. These pellets of Uranium must then be packed into Zircalloy Rods which, when delivered far and wide to Nuclear Power Plants both domestic and foreign, are lowered into position at the center of the reactor core, ready for action.

The industry calls all these dirty, laborious, ingenious and terribly terribly expensive processes described above, "the front end" of the Nuclear Fuel Cycle. Most, or all of the expense of "the front end" is paid for with American taxpayer money. The Nuclear Industry cannot in any way operate without the "the front end." But they don't want any part of the dirt, the responsibility, the tedious labor and expense of dealing with it. They prefer to have their fuel delivered to them by the American taxpayer. In their view, they have all they can worry about dealing with "the back end."

TEN

AN IMMORTAL ELEMENT THAT GLOWS AND GROWS

WORLD WAR II DEMANDED the creation of a new and terrible explosive weapon that would bring the war to an end, with victory for the side that was defending civilization. The barbarity and insane nature of the German leadership frightened business, the banks, governments, and the peoples of the world.

Nothing was spared to bring about the manufacture of the first atom bomb at Los Alamos, New Mexico, under the brilliant administrative and scientific leadership of Robert Oppenheimer, a physics professor from California, who was able to marshal the best brains available at the time, to work on this special military project.

Until 1940 P=L=U=T=O=N=I=U=M was unknown.

There are only very rare traces of it in Nature, so as a natural element, it hardly may be said to exist at all. It was first separated and identified in 1941 at the University of California at Berkeley.

What the scientists at Los Alamos needed, was a single element, a "trigger," capable of setting off a chain reaction in a bomb, a trigger capable of releasing a shattering explosion of energy millions of times more powerful than any previously known chemical explosion.

Only very rare substances can trigger an atomic explosion. Uranium-235 can do it, but it was not easily available in the 1940's. Meanwhile, experiments showed that P=L=U=T=O=N=I=U=M could be extracted from readily available "ordinary" Uranium. It could be reduced by chemical processes. In August 1945, both substances (Uranium-235 and P=L=U=T=O=N=I=U=M) were used in the atomic bombs dropped on Japan: Uranium-235 set off the bomb over Hiroshima, and P=L=U=T=O=N-=I=U=M triggered the bomb dropped on Nagasaki three days later.

P=L=U=T=O-N=I=U=M has special value to the military, and to the commercial Nuclear Industry. It is a man-made element which is created in every Nuclear Reactor. It is a by-product of Nuclear Fission, as unused Uranium is automatically turned into P=L=U=T=O-=N=I=U=M as a "waste product." It can then be separated out by chemicals and mixed with *ordinary* Uranium to make new fuel for the next round of Nuclear Fission.

The Federal Government does everything it can to promote the manufacture of P=L=U=T=O-N=I=U=M because, as we have seen, it is the necessary trigger ingredient in the Atom Bomb.

The Nuclear Industry has yet another use for P=L=U=T=O-N=I=U=M as "the ultimate and final energy source." They dream of building a new kind of Nuclear Reactor, called a "fast breeder" *which burns P=L=U-=T=O=N=I=U=M while producing more P=L=U=T=O=N-=I=U=M than it burns!* The only trouble with that dream is that "fast breeders" have a mind of their own and tend to explode uncontrollably.

Thus were they married: the military, and the civilian utility corporations who operate Nuclear Power Plants. They were made man and wife: the military could have all the P=L=U=T=O-N=I=U=M triggers it lusted after, produced as a convenient "waste product" of the plants. The more power plants there were, the more P=L=U=T=O-N=I=U=M could be generated.

Thus also was born the extraordinary secrecy surrounding Nuclear Energy. The Utility Corporations were pledged to feed the military. The production of P=L=U=T=O-N=I=U=M goes on, on an enormous scale. And P=L=U=T=O-N=I=U=M, the single most powerful form of pollutive death in existence—an element that can only be "remote handled" because flesh is harmed on contact with it, has been gathering and growing in ever increasing abundance. If the Nuclear Utilities and the military have their way with us, the Earth will *forever* harbor an immortal element that glows and grows, with its own agenda of filling up every crevice on Earth.

P=L=U=T=O-N=I=U=M has a half-life of 24,400

years. One one millionth (1/1,000,000) of a gram will cause incurable lung cancer. For every pound of usable P=L=U-=T=O-N=I=U=M that is chemically extracted, there are 200 pounds of radioactive waste materials generated, and thousands upon thousands of gallons of red-hot radioactive coolant, water, and solvents with no place to ever be safely disposed of where humans will not be harmed.

ELEVEN

SOMEWHERE ELSE, ON SOMEBODY ELSE

NUCLEAR REACTORS DO NOT LAST FOREVER, and they are the very heart of every Nuclear Plant. At the core of the reactor are the fuel rods. These are rods packed with half-inch long Uranium pellets. The rods are packed together in 14 foot columns, and each column of rods can produce energy for up to four years.

After perhaps three or four years, the now "spent" fuel rods must be removed, and new rods put in their place so Nuclear Fission can resume. The problem the utility corporations would like to settle to their satisfaction is: how to get rid of the spent rods, get them out of their plant, and forever off their property. In other words, they want desperately to dump a terrible radioactive problem *somewhere else, on somebody else.*

Spent rods are intensely intensely radioactive. Humans cannot go anywhere near them. No living thing may breathe in their presence, or ever touch them. They are critically *hot* in temperature. Spent rods must be soaked in a boric acid solution tank for up to six months or more just to cool them down enough so they can be moved.

The men running the giant utility corporations feel it is their right and their duty to dump their waste, their rods, their transuranics, their barrels of contaminated lubricants, their radioactive waters, all their contaminated machine parts—*everything*—in another part of the country where the population doesn't know how dangerous they are. They also feel that they will draw less flack to themselves from states with smaller populations, that consequently have less voice in Congress.

In their thinking, the Western States of the United States make the best targets for waste. There is plenty of beautiful open land there, lots of grassy fields and fresh green forests, and snow-covered mountains, and when these utility folks see all that beauty, they don't ever see the beauty of God's green Earth—they don't ever see what God and Nature have wrought. They see only carved out tunnels in the earth, sinkholes for oily contaminants. They see their filthy wastes increasing daily, and they want them moved out by the Federal Government and deposited *somewhere,* in somebody else's backyard, not their own. When they see a really beautiful place, they just can't wait to rig up a cyclone fence, and enclose a Nuclear Dump—preferably in a state where the population doesn't appear to be all that knowledgeable, and where the possibility of having the

Federal Government dedicate that land to them forever, for waste forever, seems passionately desirable. And *forever,* in terms of P=L=U=T=O=N=I=U=M, means just that. P=L=U=T=O=N=I=U=M is indestructible. It lives on long and long after we and our children are gone. It remains powerful, aggressive and alive for all of time. 240,000 years is one estimate of its poisonous energy.

"In their thinking, the Western States of the United States make the best targets for waste."

TWELVE

ROCKY FLATS: "A CREDIT TO THE COMMUNITY"

MILLING P=L=U=T=O=N=I=U=M is probably the dirtiest, most dangerous business ever licensed in America. Since 1951, Rocky Flats, Colorado, has been the site of the milling and manufacture of P=L=U=T=O-=N=I=U=M "buttons." It is a very large industrial foundry, owned by the Federal Government, and planted a mere 16 miles northwest of the biggest city in Colorado, which is Denver. The plant's primary mission is to smelt P=L=U=T=O-N=I=U=M, separate it to the point where it will set off an atomic explosion, and then mold it into "triggers" for Nuclear Bombs. The triggers are then shipped to a plant near Amarillo, Texas, where they are assembled into weapons. When the P=L=U=T=O-=N=I=U=M trigger explodes in an atomic blast, it sets off a hydrogen explosion.

On March 23, 1951, the *Denver Post* headline read, "THERE'S GOOD NEWS TODAY. U.S. TO BUILD $45 MILLION A-PLANT NEAR DENVER." The Atomic Energy Commission (daddy of the DOE) went to the greatest lengths to assure Denver residents that the plant would not only provide new jobs and be a credit to the community in every way, but "the plant will offer no threat whatever to the health and safety of residents of nearby communities, and that workers on the project will be safer than downtown office workers who have to cross busy streets on their way to lunch."

Following this single outburst of communication to the locals, the cloak of secrecy was dropped over Rocky Flats. Officials of the Atomic Energy Commission would only say further, that Rocky Flats would be "a secret processing production plant which will handle radioactive material."

The public was never to know what really went on at the plant, until June 15, 1957 (six years after operations began), when the AEC was forced to admit that a P=L=U=T=O-N=I=U=M explosion had occurred, in which at least two workers were very seriously injured. This unexplained explosion was followed by more accidents, many unmonitored leaks and spills, and the build-up of immense quantities of radioactive wastes, both liquid and solid.

Later, in 1989, the nearby communities that had eagerly welcomed Rocky Flats as both an economic boon and the fulfilling of a patriotic duty, were digging ditches around their drinking water supplies, as streams of yellow toxic waste flowed out of the plant and out into the countryside. A secret memo issued at the plant reported an average of 32 "contamination incidents" every month, ranging from the inhalation of P=L=U=T=O-=N=I=U=M fumes, to skin contact with radiation releases. In addition, the memo reported an average of "more than two nuclear criticality infractions" a month. Such incidents can lead to the uncontrolled fission of P=L=U=T=O-N=I=U=M, which can result in the release of large amounts of poisonous radiation into the atmosphere.

Who were the people hired to work at the plant, and how did they benefit from the employment opportunities, touted as the chief advantage to the community of having the plant in their area in the first place?

Let's look at the life of one worker, Don Gabel. Don was a fry cook working in a Denver restaurant, when he heard about hiring at the plant. He applied for a job that paid $8.35 an hour, with the possibility of earning more, for certain kinds of work, which they would tell him about later. Don had a young wife named Kay, three small children, a house and a mortgage. He wasn't doing too well as a fry-cook. He was an earnest Christian, and a patriotic American. He had heard that the work at the plant had something to do with "national defense." He was eager both to better himself, and to do his part for his country.

Don was given three days of training in the "remote-handling" of molten P=L=U=T=O-N=I=U=M. He was then told to sit in front of a "glove box," slip his hands into the permanently attached canvas gloves that are lined with lead, look through the triple-paned glass "window" of the box, and mold P=L=U=T=O-N=I=U=M "buttons" with his fingers. He then had to stack the "buttons" off to one side, as far as he could reach to make a pile. When the pile was high enough, he had to go behind the box, remove the molded buttons, and pack them for shipment. Because of the radiation danger involved in this activity, Don was paid 15 cents an hour extra in "hot pay." Buttons would inevitably fall to the floor as they were packed. When that happened, he was told to pick them up and put them into the shipment stacks, using an issue of "surgical gloves" he was given for that purpose.

Early on, Don began to suffer headaches, for which he was given aspirin. When he touched an especially "hot" piece of equipment, radiation detectors would reveal that he had been bombarded with neutrons. When that happened, he was told to go to the medical office, where technicians would scrub him down with Clorox.

Soon, the headaches became chronic and soon afterwards, Don suffered a seizure and lost consciousness. X-rays revealed that Don was harboring a large tumor in his brain. It had to be removed if he was to live.

The tumor operation left Don's head cruelly misshapen. The surgeons had to remove a portion of his skull, to get at the growth.

It was clear that Don had suffered radiation poisoning. He now realized that he had been cruelly used by the Department of Energy and their contractors, Dow Chemical, and Rockwell International.

With nearly 6,000 jobs and a payroll of $220 million, Rocky Flats is the state's eighth largest employer. The supplies it buys pours another $59 million into the economy.

Steady pay, good benefits, and the need for job security kept Don at Rocky Flats. He worked there from 1970, until his death in 1987. He had worked at the plant for about 17 years. He died at the early age of 40.

Before his death, Don spoke out bitterly about conditions at the plant. There was no ventilation for the workers, so a solid steel fire door was drilled through to admit air. Smokestacks vented P=L=U=T=O=N=I=U=M into the atmosphere from the smelting process. Steel exhaust pipes leading from furnaces used to smelt P=L=U=T=O=N=I=U=M were suspended right through the factory, into the area where workers spent eight hours a day. Don recalled that at his work station, an exhaust pipe lay six inches away from his head. When he asked if this was a danger, his supervisor made light of it, saying that he was entirely safe because the pipe was "nowhere near his stomach."

The Colorado State Appellate Court, in June 1987, declared that Don's death was due to radiation poisoning, but his widow, Kay, was denied any compensation for Don's death. Rockwell International had been made exempt by law from any and all suits brought against them either by workers or the general public. They were free of all responsibility for any mismanagement at the plant.

The Justice Department instructed the DOE not to comment on conditions at Rocky Flats. As far as the DOE was concerned, what they were doing, no matter how badly they were doing it, was done in the name of "National Defense."

In 1989, the FBI found the plant to be operating

"in poor condition generally in terms of environmental compliance." In other words, the DOE had been breaking the law for years, and had been putting the health and welfare of the general population at risk.

All this while, the DOE was awarding Rockwell millions of dollars in "performance bonuses for excellent management"—an estimated $41 million in all was given them over the years.

Now Rocky Flats is due to be closed down.[6] It may take as long as 20 years to be phased out. According to government estimates, it will take as long as 30 years, and untold billions to clean up the plant and the surrounding area. All this expenditure will come from the American taxpayer.

Meanwhile, the damage is still being evaluated. Ground water has been irrevocably contaminated, and brain cancer is four times higher in the Denver area, than in the nation at large.

Rocky Flats' veil of secrecy has finally been torn down. The damage it has done, however, will remain for generations to come.

---

[6] The question of the opening or closing of Rocky Flats is a murky one. The plant was admittedly so infected with P=L=U=T=O=N=I=U=M that the DOE announced in 1989 that it would have to be closed. Then, on June 20, 1990, the DOE announced that it was building a new facility on site, to be called "Building 371." This new facility would allow the DOE to resume P=L=U=T=O= N=I=U=M smelting operations "for a period of 30 years." (At the end of the 30 year period, the new building would presumably also have to be abandoned, because it would have become too dirty to work in.)

The creation of Building 371 is opposed by Colorado's Governor Romer, Colorado Senator Tim Worth, and Colorado Congresspersons David Skaggs and Pat Schroeder, as well as the people living in Boulder and in the areas surrounding Rocky Flats.

THIRTEEN

MOLTEN URANIUM: A FIERY ROCKET

GETTING THE "FACTS" ON NUCLEAR—Nuclear anything—is not an easy matter. The Nuclear business began at Los Alamos during World War II as a deep military secret. Ever since, they have decided to carry on in this tradition. It seemed necessary back then, to hide from everyone—the Germans, the Japanese, the Russians, our Allies, our own people—that our engineers and scientists were looking for the Atom Bomb. "LOOSE LIPS SINK SHIPS" was a popular poster you saw hanging at all army installations, Civilian Defense Headquarters, even factories. There were, after all, German spies behind every bush.

Today, it is still very difficult to overcome the veil of secrecy the Nuclear Business employs to shield its activities. Everything about Nuclear is hidden, arcane, covered up by jargon, and unfortunately, often coated over with a generous frosting of lies.

The Nuclear Industry, the Government, the Military, have "incidents" they feel they must cover up, hide from citizens. Researchers on nuclear information are all agreed, that facts are very often denied them. But blatant error and even lies are often brought to light by the people the cover-up has harmed in one way or another. Researchers generally approach the whole Nuclear Question, with a healthy dose of skepticism.

Still, certain facts have emerged. Here are a few, presented in no particular order, except that they are all rather interesting:

—Every operating Nuclear Plant produces up to 500 pounds of P=L=U=T=O=N=I=U=M a year.

—The U.S. has 112 operating Nuclear Plants. That brings the amount of P=L=U=T=O=N=I-=U=M produced in the U.S. alone, to 56,000 pounds a year.

—Consider again that 1/1,000,000th of a gram of P=L=U=T=O=N=I=U=M will cause lung cancer if ingested through the nose or mouth.

—It's worth repeating, that for every pound of usable P=L=U=T=O=N=I=U=M extracted from the waste produced by fission, there are 200 pounds of highly radioactive P=L=U=T=O= N=I=U=M-soaked waste or garbage generated that has to be very carefully isolated, or it will kill. If our multiplication is correct, what in the world can we expect to do with 11,200,000 pounds of radioactive waste a year? Who, in their good common sense, would want it set out on the ground near them? And would you want your neighbors to be forced by the government to accept this terrible burden of death dealing product?

—The Nuclear Industry says dumping is critical to their business. They say they must move out the waste generated daily at their plants in order to make room for more waste to come.

—Nuclear Reactor Plants last only about 40 years, because they accumulate radioactivity so deeply around themselves, that they must be abandoned. Radioactivity permeates the plant—every room, every pump, every tank, every pipe. The Plants become so "hot" that they can no longer be repaired or serviced. That means that they have to be isolated, by covering them over, to try to shut them out of the environment. They will then lie "hot" and be an active center of contained (hopefully) radiation for 1.5 million years. Can you imagine those generations following us, coming upon mysterious domes of cement, not knowing what they are? The signs will all be faded out by the sun and the elements. The cyclone fence will be blown down and gone. They might conceivably build a city next to the dome, and dig for water. What a heritage we are leaving to our children's children and to their innocent children's children.

"Can you imagine those generations following us, coming upon mysterious domes of cement, not knowing what they are?"

—A "LOCA" is a "Loss Of Coolant Accident." A "LOCA" can happen at any time, at any Nuclear installation. At 5000° F. everything in the plant melts down—fuel rods melt and split open, steel containment walls are breached, vessels containing radioactive fluids crack apart and flood out, and when Uranium turns molten, it will pass through the bottom of a reinforced concrete floor and burn its way 50 feet into the Earth. This is called "The China Syndrome." Everything's headed South toward China, like a fiery rocket.

—When this fireball of molten Uranium hits ground water, steam is produced, and that's what can cause an uncontainable explosion. Everything above ground will blow and P=L=U=T=O=N=I=U=M radiation is spewed out widely and indiscriminately over a very grand perimeter.

—The Nuclear Industry is at constant pains to tell us that a "LOCA" will NEVER happen. NEVER? They have designed a five foot thick steel and concrete floor and walls to *contain* the meltdown. However, it has never yet been tested. Remember the *Titanic?* It was a freshly designed luxury ocean liner advertised as "unsinkable."

—In 1972 the U.S. Government commissioned the Rasmussin Report. It was a $4 million study of the problem of meltdown. Professor Rasmussin of the Massachusetts Institute of Technology was encouraged by the Government to *minimize* the dangers inherent in a typical meltdown. The Union of Concerned Scientists criticized the report as unrealistic. The American Physical Society (of Physicists) faulted the report for not paying attention to low-level radiation as well as to P=L=U=T=O=N=I=U=M related high-level radiation.

—Rasmussin was forced to revise his report. The Government (through the report) admitted the following, in the case of a meltdown:

- 3,300 people in the surrounding area would die immediately—
- 45,000 people would be burned and dosed, but would not die immediately. In their suffering they would linger for an indeterminate period of time—
- 45,000 people would develop various P=L=U=T=O=N=I=U=M-related cancers—
- 5,000 birth defects would surface, ranging from disfigurement, to mental retardation—

—The Nuclear Industry has always been at pains to say that Nuclear Power is the cheapest power on Earth. It isn't, when you consider that the American Taxpayer has allowed the "socialization" of the Nuclear Industry. The taxpayer pays to help build every Nuclear Plant, pays all the heavy waste storage costs, pays for the fuel, pays for all clean-ups, and pays for all research and development.

—All in all, Nuclear Power is not only the most dangerous power, but clearly the most expensive. According to Rasmussin, it would take a minimum of $14 billion to clean up a meltdown. But that was in 1979 dollars, when the report was first published.

—According to Rasmussin, the chances of a meltdown occurring are one in every 200 years. However, Rasmussin figured on only 100 operating reactors in 1972. We now have 112 operating reactors, in 1992, in the U.S. alone. His "200 years" was a theoretical stab in the dark, designed more to quiet fears, than to be reliably reassuring.

FOURTEEN

ACCIDENTS WILL HAPPEN

NO SYSTEM IS FAIL-SAFE. Nuclear Plants are built carefully so as to minimize "accidents." Systems are never perfect, because they are built and managed by humans. Many "accidents" have already occurred and are still occurring. But when we are dealing with irretrievable radioactivity, then every accident or release adds fatal damage and insult to an environment already overloaded with many threatening contaminants.

In March of 1979 the Three Mile Island "accident" took place in Pennsylvania. The Nuclear Regulatory Commission sent in a team of experts who were not able to examine the red-hot reactor and determine its condition, since to approach it would mean death within thirty seconds.

Thousands of gallons of radioactive water had already been released into the nearby Susquehanna River. When a river is made radioactive, it means that somewhere down the line, lots of folks will get to drink cancerous water.

Metropolitan Edison, the utility corporation that owned and operated the plant admitted that their personnel did not have enough training to read the controls correctly.

The plant was sloppily managed. Great stalactite mineral deposits had been allowed to build up on valves and pipes, effectively locking the controls.

The same valves were locked at a similar plant in Ohio, by the build-up of leaking boron stalactites. That occurred a year earlier.

Metropolitan Edison instituted a $4 billion suit against the NRC, and a $500 million suit against the valve manufacturer. Metropolitan Edison was simply not ready to accept any responsibility for the accident themselves.

In an earlier incident in 1975, the cooling system burned out at Brown's Ferry Nuclear Plant near Decatur, Alabama. Workers had held a candle up to the rubber packing around an air leak to see if the flame would flicker. There was an immediate explosion and a terrible fire broke out that completely closed down the plant.

In October of 1978 at the Federal Government Laboratory in Idaho, uranium began to go critical. The plant supervisor missed reading the dials because he was watching the World Series on a little portable TV that he had brought to the plant from home.

In 1966 there was a breeder reactor that went critical in the Enrico Fermi Reactor Plant near Detroit. The core began to melt. Now, a breeder reactor can go off like a nuclear bomb. Fortunately, they were able to close down the reactor just before it exploded. They later found that a piece of scrap metal had fallen into a coolant duct and closed it, which allowed the core to overheat.

The Indian Point Plant in New York State, owned by Consolidated Edison, suddenly began to flood. Radioactive water began to drown the reactor. That created thousands of gallons of lethal poison that would somehow have to be disposed of. Officials at the utility company called it "a pumping problem."

Nuclear terrorists could bring about a critical meltdown almost at will. All they need do is cut off the supply of coolant to any reactor. They can do this by simply turning the wheel that shuts down the coolant valve—that is, if the valve isn't frozen in place by dripping boron.

From April 1978 to September 1979 a four inch diameter pipe to the outside of a plant was accidently left open for seventeen months. This happened at a Michigan Reactor Plant. The pipe lead straight out of the containment building which is designed to contain radioactivity in case of a meltdown. If an accident had occurred during that time period, the population around the plant would have been irretrievably irradiated.

At an Oregon Nuclear Plant, the security guards were operating a drug ring. At a uranium processing plant in Tennessee, they "lost" twenty-two pounds of bomb-grade uranium.

What has to be borne in mind in dealing with Nuclear on a day-to-day basis is that a major accident would not only kill and maim, but such an event would infect the food chain for centuries to come. And it would do so without warning. We ourselves would not be able to detect it or to know that it had happened, since radiation is undetectable by human senses.

FIFTEEN

SPREADING NUCLEAR AROUND

THERE ARE SO MANY IRONIES connected with Nuclear that it reminds us again of the sister science of astrology, namely Alchemy. Alchemy began in ancient Egypt or China as an attempt to realize the dream of changing one substance into another. It's chief ambition was to make money by turning base metals, say, iron or copper, into gold. That would make the Alchemist rich beyond belief. He could then command and create empires.

By the 12th century, Alchemy had migrated to Europe where it developed its own vocabulary and secrets. These secrets were often connected to magic and symbology. Alchemists developed a mysterious brotherhood, and a jargon to go with it, that was expressly cryptic, to confine information within the brotherhood. It was a secret society that really believed that by experimentation, by mixing elements together, it could finally be rewarded with huge profits. Since gold was the universal currency, then as now, they could finally be wealthy enough to buy the world.

The Nuclear Industry appears to be the Son of Alchemy. It is a secret society that jealously guards its secrets from the public. It has been allowed to do so because of the military applications of P=L=U=T=O=N-=I=U=M. After all, P=L=U=T=O=N=I=U=M has all the earmarks of the Alchemist's dream. It has become valuable by processing a base element, Uranium, into a form that is equal in value to gold. It is an accidental by-product of fission, and so long as Nuclear bombs are the currency, it can be traded for very great sums of money.

The Nuclear brotherhood is hidden. Who are the people buying and selling P=L=U=T=O=N=I=U=M? The secret of the creation of P=L=U=T=O=N=I=U=M is a well-guarded one. The general public has been told only what the giant corporations who buy, sell and fabricate P=L=U=T=O=N=I=U=M will allow it to know, and they guard their information jealously.

Alchemy was never a "science," but the Nuclear Industry has aggressively attempted to wrap itself in the protective cloak of "science." Science, divorced from humanity cannot be science at all. It is more like voodoo. And voodoo can-do all kinds of mischief.

It is not science, if it endangers people's health. It cannot be science if it uses people exactly as if they were machines. When a machine breaks down in their industry, they buy a new one. (Sure, it can't be repaired; it's burning hot with radioactivity.) It can't be science if it employs the principle of "acceptable risk." "Acceptable Risk" means that the population downwind of every P=L=U=T=O=N=I=U=M producing plant will develop cancer. So what? "Acceptable Risk" means that poor people in poor neighborhoods can be subjected to often grotesque forms of radiation poisoning. And now it's not only poor people.

In its zeal to become even more wealthy and powerful, the new Alchemy has boldly overstepped its boundaries. It now wants to harm even rich people in rich communities. The new Alchemy has made plans to transport P=L=U=T=O=N=I=U=M down the main streets of selected cities and towns in the nation. That is a bold step! It has built giant flat bed trucks capable of hauling steel cannisters filled to the brim with P=L=U=T=O=N=I=U=M-infected waste. It wants to truck these wastes down main streets, past schools and churches, past businesses, past homes. If a cannister should ever rupture, should a drunk driver strike and overturn one of these monster trucks, then the wind would carry P=L=U=T=O=N=I=U=M to every housetop, dump granules into yards, blow contaminated particles into every shop and home. Since it has never yet happened, no known method of clean-up has ever been demonstrated a success.

The "Alchemists" have said, "It's OK. The police and firemen will take care of everything." But the police and firemen entering a highway spill-zone smoking with P=L=U=T=O=N=I=U=M, would be committing suicide. The Alchemists know better than to send in their *own* sons, but are unfeeling enough to say to the local municipalities, "It's your patriotic duty to take care of it." Meanwhile, the giant corporation trades its stock over the counter on Wall Street, while our police and firemen, husbands, wives and parents all, are asked to make the supreme sacrifice.

Where are the trucks going, anyway? The trucks have two drivers: one drives while the other rides shotgun.

They are going to a cow pasture in the great Southwest where the giant Alchemical Corporation has already dug out a vast salt mine in which to store the cannisters. In the mine, which is carved out of a salt deposit at the bottom of an ancient sea-bed, salt water drips and runs

"[The waste is] going to a cow pasture in the great Southwest where the giant Alchemical Corporation has already dug out a vast salt mine in which to store the cannisters."

like underground rivulets. When the cannisters sit in the salt water, they begin to rust and corrode, and the salt water forms a slurry from what is leaking from the cans, and the now red-hot salty P=L=U=T=O=N=I=U=M-acid begins to run toward the underground aquifers that the people here have drilled into for their drinking water, and for watering their cattle. Very soon, the people for miles and states around notice they are subject to strange cancers. American children are born deformed and mentally deficient. And in New York, a beautiful and juicy steak is brought to the table, which was recently carved out of an animal that drank contaminated water in New Mexico.

SIXTEEN

NEVER SAY DIE

P=L=U=T=O=N=I=U=M IS AT THE HEART of all the effort going into military defense expenditure. It is the centerpiece driving all military planning and execution. It is more valuable than gold and far more expensive. It is the currency of nations, not individuals, because P=L=U=T=O=N=I=U=M is the key to world power and domination.

Nuclear Bombs feed on P=L=U=T=O=N=I=U=M. It is their bread. They would starve without it. A Nuclear Bomb is a lifeless piece of stone-cold hardware, without its trigger. And its trigger is P=L=U=T=O=N=I=U=M. A Nuclear Bomb can only explode and spread the ultimate destruction and poison it is designed for, when it is sparked by P=L=U=T=O=N=I=U=M, the Devil's own metal.

For that reason alone, for the by-product they produce, which is P=L=U=T=O=N=I=U=M, governments are eager to bankrupt themselves in order to build Nuclear Plants. In the U.S., Nuclear Plants are managed by giant Utilities, who were persuaded long ago that they would always remain a certain source of profit, since the huge debt created when the plants are built, would be underwritten by the public. Nuclear Plants produce only a trickle of electricity, compared to coal-fired or oil-fired plants. They produce so little because they break down so regularly. They are extremely difficult to manage, and have to be closed down when there is a stoppage of any sort, or when the central core has to be replaced, or even for routine maintenance.

The Utility barons have tried to persuade us for over forty years that Nuclear is the wave of the future, that Nuclear is "clean," and that it is inevitable.

What the barons never mention is that Nuclear Power Plants, besides producing P=L=U=T=O=N=I=U=M, produce very great amounts of P=L=U=T=O=N=I=U=M-soaked waste, and that waste is, as we have said before —*Immortal.*

Not only is it *immortal,* and not only is P=L=U-=T=O=N=I=U=M-infected waste growing and spreading, but it can produce poisonous gas. You know how uncomfortable gas is. Gas inside us wants to vent to the outside but cannot be vented in polite society, and for good reason. Right?

Not so with Nuclear. Nothing about Nuclear is polite or social.

The Giant Corporation that builds Nuclear Power Plants and sells them to the world, is also contracted by the U.S. Government to carry off the waste that is piling up every day at every Nuclear Power Plant. Further, the Corporation has been awarded the mission to truck P=L=U=T=O=N=I=U=M through 24 states, and that means through hundreds of cities, towns and counties of this nation, to a supposed final resting place in Carlsbad, New Mexico.

Nine miles of galleries have been drilled into the salt there, at a cost of $800,000,000. The Corporation has, for the last 14 years, told everybody that the dump is "scientific."

What is meant by "scientific," no one knows. What they have in mind is that they are being paid huge sums of public money to fulfill the mission given them by the Federal Government to haul away all the trash being created at defense facilities and nuclear power plants. "Scientific" means "profitable," to the giant trash corporation and to the power industry.

The manager at The WIPP dump, which is the "National Monumental Trash Site," and the prototype for more, said recently, "No stone will go unturned." What did he mean by that cryptic remark?[7]

At WIPP, P=L=U=T=O=N=I=U=M-soaked waste is to be stored in 55-gallon steel drums, having been put to rest by lifting machinery especially created and designed

---

[7] WIPP means Waste Isolation Pilot Plant.

by The Corporation. A human being has been hired by the Giant Corporation, put into a mask and helmet, and armed with lead-lined gloves, has been trained to drive the lifts.

The caverns receiving the waste were drilled 2,150 feet below the earth.

The Corporation is convinced that buried that far *out of sight*, no one will remember it is there. Their motto is, "Out of sight, out of mind."

Gas is generated when the drums corrode in salt water and the water mixes with the organic material and chemicals in the drums. The problem with gas is that it can build up massive pressures under the salt dome within a 60 year period. That pressure will crack the walls of the salt deposit galleries, and allow active radioactivity to leak out into the environment. The chambers are already cracked, with large fractures running out beyond the site. As the gas builds, it finds the fractures and enlarges them, allowing salt water to run through and find a downhill egress into the underground drinking water system.

Gas also tends to gather and explode, and when it blows, it will carry with it a vast quantity of P=L=U-=T=O=N=I=U=M which will enter the atmosphere and be borne aloft by the jet stream. It could be a Son of Chernobyl in the making.

The Giant Corporation is always ready with either "solutions" or excuses. As an alternative, it now wants to *burn* P=L=U=T=O=N=I=U=M in the atmosphere, to reduce its volume by incineration. It also has plans to put the trash into plastic barrels so there won't be any rust. It also wants to *vent* the gas into the upper atmosphere to be carried aloft, South to Texas, East to Chicago, West to Los Angeles and North to Denver. Or, it asks—Why not water down the trash so it isn't so volatile and dense? That way, it may take as long as 120 years to explode, instead of 60. Then again, chemicals might be added to the volatile trash to "absorb" the gas, like the TV commercial that settles a cartoon stomach after a really bad meal. And how about encasing P=L=U=T=O=N-=I=U=M in concrete *before* putting it into barrels? Or surrounding it with melted glass? If all else fails, why not pipe the gas into further chambers, seal the chambers shut, and keep our fingers crossed?

All of these "solutions," in their view, are densely "scientific." But, their definition of "science" really seems to mean: let's do *something*; let's not just sit around; let's get the bonus the Government has promised us for doing the kind of job we contracted to do, no matter what!

In their view, the *simplest* solution is just not profitable to themselves. The simple solution is, barring fundamental answers: leave the damned P=L=U=T=O=N-=I=U=M-soaked infection where it is produced, at the Nuclear and Weapons Plants that produced it. (Better yet, stop producing it.) Why put 24 states and millions of our own people at risk by moving P=L=U=T=O=N-=I=U=M around and burying it where it will simply never say "die."

SEVENTEEN

THE GREAT TAXPAYER NUCLEAR GIVE-AWAY

FACTS, AS SUCH, DON'T GIVE US A "FEEL" for the problem we face every day, of being irradiated. But "facts" by themselves, can be very interesting. They can be given out straight, so you, the reader, can think your own thoughts about them and come up with your own conclusions, deciding where your own best interests lie. We like facts used that way, where you get to sort them out, put them together yourself.

Here are some. They don't necessarily lie in any special order:

—Congress set up the Atomic Energy Act in 1954.

—It set up the Atomic Energy Commission (AEC) as the licensing authority for all Atomic Power Plants.

—The mission of the AEC was two-fold: it licensed the industry, and it *promoted* Nuclear Power for "peaceful purposes."

—The first chairman of the AEC was Glenn T. Seaborg. Seaborg could not bring himself to believe that a nuclear accident would ever happen.

—Because of Seaborg's "optimism" (some say blindness) Nuclear Power Plants were allowed to be built near great cities, near great and intense population centers. He allowed the Indian Point Plant in New York to be built 20 miles from Bronx, New York, 40 miles from Times Square. More than 20 million people live within a 50 mile radius of that plant.

—He argued for the Zion Plant Reactors in Illinois to be built 41 miles from Chicago, and only 40 miles from Milwaukee.

—Seaborg's buddies were the Utility Corporations' managers, and they wanted their plants built as close to their customers as they could get them. That would cut down wiring, cable costs and maintenance. They didn't give any thought to safety.

—The Virginia Electric and Power Company was permitted to build its plant, "The North Anna," 50 miles from Richmond, directly over a known earthquake fault. The Utility Company went to great lengths to cover up the research of the geologists about the danger.

—The AEC, always warm and friendly to the Utilities, gave surprisingly little thought to population safety. They knew, for example, in the 1960's, that the Emergency Core Cooling System (ECCS) in use at the time, was faulty. It was supposed to prevent "meltdowns" but it was not dependable because it was inadequate for the job. They approved it anyway. The Federation of American Scientists warned the AEC of the danger inherent in that decision.

—The AEC did not require older plants to update their deficiencies. The AEC operated on the principle that Nuclear Power was inherently safe, and there would never be any problem—*ever!*

—Plants are not required to report either radioactive releases into the air, or infected water releases into rivers and streams.

—For 35 years P=L=U=T=O=N=I=U=M-contaminated waste has been piling up. Every Nuclear Reactor generates 30 tons of toxic

P=L=U=T=O=N=I=U=M mixed-waste a year (all of it highly lethal). We have 112 plants in the U.S. producing 3,360 tons of immortal waste each year, waste that will never die. Nuclear garbage remains alive, active and volatile for centuries and centuries to come, no matter where or how it is stored. It is an indestructible kinetic waste that seeks out air and water in the environment. It is so powerful a substance that only one particle of a gram causes lingering and tortured death.

—In 1979 there were 22 regular inspectors assigned to inspect all existing Nuclear facilities. The Utility Companies were always told ahead of time when to expect the inspectors.

—If a plant inspector suggested a safety improvement, the Utilities were never required to follow up on the suggestion.

—There are no evacuation plans for populations, in case of a Nuclear disaster.

—In 1974 the now familiar "Department of Energy" (DOE) was created by Congress. It replaced the old AEC. However, the same faces appeared again in the driver seats, only with new titles and higher salaries.

—Congress had simply created a "new" Nuclear

Regulatory Commission, calling it the Department of Energy.

—Robert Pollard was on the staff of the old NRC, and then left the DOE in 1976. He joined the Union of Concerned Scientists. He said, the NRC knows very well about existing problems and dangers at the plants, but hasn't the slightest intention of fixing them. Fixing old plants, for example, could cost the Utility Corporations $10-50 million for each plant. The Utilities have no intention of spending that much money and not improving their cash flow.

—The Federal Government has spent well over $50 Billion to build up the Nuclear Power Industry.

—The Federal Government provides free Research and Development information to all privately owned Utilities.

—Our tax dollars support research labs like Oak Ridge in Tennessee, the National Reactor Testing Station in Idaho, and all the facilities at Los Alamos National Laboratory.

—No private company has ever wanted to come forward to provide Uranium-Enrichment Plant Facilities. All three such plants in the U.S. are government owned, managed by private contractors. Enriched-Uranium is sold at cost to the privately owned Utilities.

—No Utility is willing to pay for any clean-up of P=L=U=T=O=N=I=U=M infection caused by themselves. They are exempt by law from doing so. The Federal Government is expected to pay all clean-up costs. Utilities cause the mess; taxpayers rush in to pick up their mistakes.

—After the Three Mile Island "accident," Metropolitan Edison, the private Utility which owns the plant, called on the Federal Government to come in and clean up their plant. So we all paid $1.5 Billion for their poor management.

—The Federal Government has accepted all the costs related to P=L=U=T=O=N=I=U=M-infected waste produced every single day at Nuclear Power Plants. The Utility puts the waste in cardboard boxes or plastic bags or steel drums, places it at the back door of the plant, and then calls for the Department of Energy to come and pick up the garbage, at our expense.

—It costs $811.1 million to build a Nuclear Power Plant. It costs only $638.4 million to build a coal-fired power plant. Both produce the exact same amount of electricity.

—The Utilities don't want to build any more coal-fired plants. Why? The Public Utilities Commission (PUC) allows the Utility to charge the rate-payers (that's us) on a "cost plus" basis for Nuclear Power Plants only. That means, the Utilities can write off buildings, maintenance and equipment, then add a profit margin of their own choosing, and *voilá*—that's the cost of their electricity to us. Because "cost-plus" is a Federal Government allowance plan, the Utility can earn $38 million more in the *first year* of operation of a Nuclear Power Plant, as opposed to a coal-fired plant. It's all a "fool-the-eye" plan worked out between the Nuclear Industry and the Federal Government. Of course, we, at home and in industry are forced to foot the bill. That way, the Nuclear Power Industry can lay claim in their advertising, that they can produce "cheap electricity." They hide the overwhelming secret costs we are paying for them.

—Nuclear Power Plants can only provide 60% of the power they are dedicated to. The reason: every time anything goes wrong, the whole plant has to shut down. Even routine maintenance often requires a plant to be taken off-line.

—Coal-fired and oil-fired plants provide 70% of their rated productivity.

—When the Three Mile Island Plant in Pennsylvania almost blew up, the officials at Metropolitan Edison, the Utility owners, wanted (get this): EVERY RATE-PAYER IN AMERICA TO PAY INCREASED ELECTRIC RATES TO PAY FOR THE UTILITY'S SHARE OF THE PLANT CLEAN-UP. *They said it was America's patriotic duty to keep the plant actively producing* P=L=U=T=O=N=I=U=M *since it was a matter of "national defense."*

"The owners of Three Mile Island. . . wanted. . . every rate-payer in America to pay increased electric rates for the utility's share of the plant clean-up."

EIGHTEEN

THE "PRICE" OF NUCLEAR PROTECTION

BECAUSE OF ITS SECRECY of operation, and its dangerous and complicated technology, and because of its close association with the military, the Nuclear Power Industry has always set itself apart from both commerce and the ordinary lives of the citizens of the nation. What the industry did and how well it performed was evaluated by an agency of the government almost as secret as the industry itself. Even congress did not understand how the Nuclear business operated. The business operated beyond the normal principles of capitalism. It had nothing to do with market forces.

P=L=U=T=O=N=I=U=M is a by-product of fission, produced almost incidentally, and fission itself is nothing more than the production of extreme heat which is then piped to boilers to produce steam. It is this steam which turns the generators that lights the lights. During this process of fabrication or production, P=L=U=T=O=N=I=U=M is "left-over," and this man-made element, which is not to be found in Nature, is what is required to bring the Atomic Bomb to life.

The Nuclear Industry has always denied that its buildings, plants and processes are dangerous. At the same time it worried about what would happen to them financially if there should be a meltdown or explosion. To guard itself against the public, it invented and carried through the U.S. Congress, the so-called Price-Anderson Act of 1957.

Since no insurance company in the nation will insure clients against a "nuclear disaster" it was assumed that the Nuclear Corporation could be brought to law and made responsible for its carelessness, mistakes or disasters. Nothing is farther from the truth.

The Nuclear Utility Corporations are saved from having to pay for their mistakes by the Price-Anderson Act of 1957.

Consider this: if Three Mile Island would have had to be evacuated, it would have cost the residents and the State of Pennsylvania $3 to $7 billion, back in 1957 (and in 1957 dollar values).

But, the Price-Anderson Act protected the Utility Corporation in Pennsylvania (indeed, any Utility Corporation, anywhere) by putting a *cap* on culpability and financial responsibility of up to, but not more than $560 million. If the people of Pennsylvania, or the State of Pennsylvania sued the Utility Corporation for the damage it had done, the Utility was protected by the $560 million cap. As arranged by the Nuclear Industry, all expenses after that figure are to be borne by the citizens themselves![8]

How did such a law as Price-Anderson get through the Congress of the U.S.? The Utility Corporations were determined to protect themselves. They paid large sums of money into the election war-chests of two congressmen: Melvin Price, a Democrat from Illionis, and John B. Anderson, a Republican from the same state. Both served in the House of Representatives and thanks to the money put at their disposal, both men were re-elected.

On going back to the House, they immediately got together, as promised, to write the Price-Anderson Bill. It is probably not far from the truth to believe that the Utilities also promised other Representatives, and Senators as well, that when their elections were due, they too would have someone in their corner to help them financially, if they would vote their approval of the bill. After all, the health and vigor of the Nuclear

---

[8] In addition to a financial cap, there is also a "time limit cap" of ten years, called the Statute of Limitations. Cancers from nuclear wastes may take up to 20 years to develop. If a cancer results from a nuclear accident, the person harboring such a cancer would be unable to file a claim after the 10 year period. He or she would be entirely on their own.

Another thing: the first party to draw on any government money would be *the utility company that allowed the accident.* They are to be compensated, up front, for all costs of suits against them. Government money, which is supposed to go to accident victims will go first to utilities to pay their lawyers to challenge victim's claims, and to pay the costs of all claims for property damage filed against the utility.

Industry meant that the National Defense of the United States would be served against its enemies. A vigorous Nuclear Industry meant an assured supply of "waste"—that is, P=L=U=T=O=N=I=U=M for Nuclear Bombs.

It was Congressman Price who came up with the $560 million figure. When he was asked how he came up with that particular amount, he explained that he scribbled it on the back of an envelop in the men's room. He said it came to him "out of thin air." (More appropriately, he might have said it came to him "out of *blue* air.")

It is not only Price-Anderson that protects the Industry against the ordinary citizen, but the Federal Tax Laws have been written to favor Utilities that operate Nuclear Power Plants. The tax laws encourage the Utilities to operate their plants even if they are not safe. For example, the officials at Three Mile Island knew their plant was in a dangerous condition, and that a meltdown might occur. However, according to the tax laws, if they could keep the plant working for four consecutive months of that particular tax year, they would save $5 million in tax allowances. Safe or not, they put the plant on-line in order to fit the requirements of the tax benefit, even if it meant putting great numbers of people at risk. Plant officials were clearly thinking about the windfall.

All this means that no one is contracted to help you in the matter of a fall-out in your area. The Utility Corporations that are responsible for operating Nuclear Power Plants safely, are protected against financial ruin by Price-Anderson. Your insurance company exempts itself in case your house or apartment is no longer fit to live in. And the corporation transporting P=L=U=T=O-=N=I=U=M through your community is protected by Price-Anderson.

> Price-Anderson encourages irresponsibility at each stage of plant development and operation; and it would deny victims of a Nuclear Accident just compensation for their losses.
>
> Ted Weiss
> Democrat, New York
> House of Representatives

> Those who would choose to poison our own people in order to make Nuclear Weapons, should be asked what the weapons are supposed to protect us from.
>
> John Glenn
> Democrat, Ohio
> U.S. Senate

THE FOLLOWING IS A PRICE-ANDERSON "UPDATE"

There was a large outcry against the $560 million cap from the nuclear industry which is developing plans to expand the number of nuclear power reactors.

Joining with them were the State Governors and the Universities contracted to operate weapons laboratories. They were worried about liability. The $560 million figure was out of date. It had to be raised.

The above groups lobbied congress for a raise. As a result, Congress amended Price-Anderson in 1965, again in 1975, and again in 1988.

Now known as *The Price-Anderson Amendments Act*, it took effect when Ronald Reagan signed it into law on August 20, 1988.

The new cap is set at $7.3 billion *per disaster*. This is wholly taxpayer money which the federal government has dedicated to pay for any and all mistakes committed by the Nuclear Industry.

While it is true that each reactor must carry insurance of $63 million, this amount would be used to rebuild the reactor should it blow up.

In the case of the nuclear waste contractors, they are exempt from all liability for any disaster caused by themselves.

It is still the obligation of the individual citizen to *prove* that he has been hurt or injured in his body or his property by a nuclear accident. Often, this means resorting to law, to class-action suits, etc., against an agency of the federal government for redress of grievances.

Since each disaster now has its own cap, it is we the citizens who are held accountable for those caps in the same way that we are going to be called on to pay higher taxes to pay for the consequences of the thrift industry.

So, where will the Federal Government get all that money?

If there is any—from you!

NINETEEN

SOME NUCLEAR QUOTES AND QUIPS

As the chief staff official of the Nuclear Regulatory Commission, Victor Stello Jr., was accused, at a Senate hearing, of altering agency regulations at the request of the Utilities that operate reactors, to the detriment of safety.

[Thanks, Victor.]

[The Bush administration has named Victor Stello, Jr., as Principal Deputy Assistant for Defense Programs. November, 1989.]

"P=L=U=T=O=N=I=U=M: a profoundly

unpleasant industrial material."

—Walter C. Patterson,
The Plutonium Business.

P=L=U=T=O=N=I=U=M is fiercely toxic in minute quantities and immensely costly to produce. It is the fuel of nuclear power.

P=L=U=T=O=N=I=U=M is named after Pluto, the Greek God of the underworld.

When Uranium-238 is struck by a particle called a neutron, its nucleus is split into two. This constitutes nuclear fission.

A Chain Reaction is produced when neutrons strike still more nuclei and split them, releasing still more neutrons.

It is a chain reaction, set off or "triggered" by P=L=U=T=O=N=I=U=M, that is the essence of a Nuclear Bomb.

The same phenomenon occurs in a nuclear reactor when neutrons strike other nuclei of Uranium-235. The chain reaction keeps on going, with a consequent extremely high rise in temperature.

P=L=U=T=O=N=I=U=M, the most "profoundly unpleasant industrial material" on Earth, gathers during a chain reaction when a neutron strikes a nucleus of Uranium-238 which "swallows" it. It has then become transformed into P=L=U=T=O=N=I=U=M-239.

"P=L=U=T=O=N=I=U=M, the most 'profoundly unpleasant industrial material' on earth."

P=L=U=T=O=N=I=U=M-239 may then "swallow" other neutrons during a chain reaction and be changed into P=L=U=T=O=N=I=U=M-240, or P=L=U=T=O=N-=I=U=M-241, or P=L=U=T=O=N=I=U=M-242.

P=L=U=T=O=N=I=U=M is nothing more than chemical garbage. But, it must be separated out from unwanted Uranium and other broken fragments of split nuclei, and further fission by-products. Separating P=L=U-=T=O=N=I=U=M is an extremely demanding process, since no human being may knowingly breathe it and live.

In the 1950's the so-called Purex Process was developed at the P=L=U=T=O=N=I=U=M processing plant at Hanford, Washington, to extract P=L=U-=T=O=N=I=U=M-239 for the military. P=L=U=T=O=N-=I=U=M-240 was not wanted for Bombs because it was unreliable. As a result of this extractive process, millions and millions of gallons of radioactive liquid is produced, because P=L=U=T=O=N=I=U=M demands innumerable extractive washings. What in the world is to be done with all this burning hot water, acid, oils and organic solvents—millions and millions of gallons of it?

There was no solution for it then, and there is no solution for it now.

I am pessimistic about the human race because it is too ingenious for its own good.

—E. B. White

A white granular powder . . . had fallen like snow upon the roofs and the lawns, the fields and the streams. No witchcraft, no enemy action had silenced the rebirth of new life in this stricken world. The people had done it themselves.

—Rachel Carson

A cell . . . under the influence of radiation . . . develops a mutation that allows it to escape the controls the body normally asserts over cell division. It is able to multiply in a wild and unregulated manner. The new cells resulting from these divisions have the same ability to escape control, and in time enough such cells have accumulated to constitute a cancer.

—Rachel Carson

TWENTY

RISK AND MORE RISK

"At a typical large nuclear power plant, the reactor core consists of about 36,000 fuel rods, containing about 100 tons of Uranium. The rods are arranged vertically in the core with space between them for cooling water to flow, and for control rods, which are used to control the rate of fission, or to shut down the chain reaction entirely.

To start a chain reaction, the control rods are slowly withdrawn. Plant operators control the amount of power the reactor produces by varying the distance to which the control rods are withdrawn.

Even with precaution, the risk of

accidents that could release large amounts of radioactivity remains. To prevent the core from overheating, water must be constantly forced through the reactor vessel.

Even after the reactor has been *scrammed* [the chain reaction halted] it must continue to be cooled, because the core is so radioactive that decay heat produced by the fission products is sufficient to cause the core to overheat. Overheating could cause a *core meltdown,* an accident in which the fuel rods melt."

Hilgartner, Bell and O'Connor
Nukespeak

★ ★ ★ ★ ★

"No facilities are available to treat radioactive and mixed waste . . . [but] the Laboratory must somehow dispose of the waste.

The Laboratory has determined that incineration of chemical, radioactive and mixed wastes is an appropriate . . . method of disposal.

High temperature incineration produces a relatively innocuous substance. It does not, however, destroy radioactivity, but only reduces its physical volume. The [P=L=U=T=O=N=I=U=M] remains in the incinerator.

We are to consider that certain risks

are allowable, in the wake of solving the issue of reducing the massive volume of hazardous materials." [Italics, ours. Eds.]

Los Alamos National Laboratory
Los Alamos, New Mexico

[Eds. note.] *Los Alamos National Laboratory (LANL) plans to make* P=L=U-=T=O=N=I=U=M *incineration available to each state in the U. S., as their solution to dealing with the build-up of radioactive waste.*

★ ★ ★ ★ ★

"Department of Energy plans do nothing to resolve the very serious problem of gas, generated by drums of radioactive waste.

After the site is filled, WIPP [a plan to bury radioactive waste more than 2000 feet in the Earth] will . . . pressurize like a balloon, to the detriment of the seals designed to prevent the [P=L=U-=T=O=N=I=U=M] in the site from reaching the environment."

"Letters to the Editor"
Thomas Alexander
The Albuquerque Journal
Albuquerque, New Mexico
December 1, 1989

★ ★ ★ ★ ★

"I am alarmed to learn about the [P=L=U=T=O=N=I=U=M] processing plant to be built at Los Alamos.

. . . I have been told that such a plant is much more dangerous than even the WIPP project itself.

The project was voted $44 million last September, and that it has already received $32 million over the last two years.

Is such critical congressional legislation so closely guarded that citizens may only find out its contents some two years after its passage? . . . And this is an issue that will vitally affect our lives."

"Letters to the Editor"
Robert Shaw
The New Mexican
Santa Fe, New Mexico
December 1, 1989

TWENTY-ONE

THE GREAT P=L=U=T=O=N=I=U=M COVER-UP

AN "INCIDENT" OCCURRED in Russia during the winter of 1957-58. A military P=L=U=T=O=N=I=U=M facility blew up in the Ural mountains. An unknown number of people were killed and a greater number were immediately injured.

The Russian government moved immediately to cover-up the accident, and in this, they had the full help and cooperation of the U.S. government. It was an extraordinary effort of cooperation between governments then waging the "cold war."

For the next 19 years, all details of this extraordinary event lay hidden. The reason for this in the U.S.

was, the Nuclear Industry was convinced that their business would be hurt by public opinion in the U.S., if people learned the facts.

In 1976 (19 years after the explosion) a Russian emigré biologist of international reputation, named Zhores Medvedev, published the first details of the accident in the British journal *New Scientist*, published November 4, 1976.

He described the radioactive blanket of dust that was spread out over hundreds of square miles surrounding the facility. Thousands and thousands of people suffered radiation poisoning, while, of course, hundreds died immediately.

Milk, meat, cattle, sheep and crops had all been contaminated. More than 30 cities and villages had to be burned down. Populations were moved out to new areas. They had to abandon their homes, their equipment, their use of all property—everything—and be moved out by truck with only the clothes on their backs.

The American CIA had all these details in their files, but would release no information until forced to do so in November, 1977, due to the Freedom of Information Act. Even so, the files were released with a patchwork of deletions, attributed to claims of "National Security."

After Medvedev's article was published, the Nuclear Industry tried to discredit him. Three Los Alamos National Laboratory staff members, led by Harold Agnew, President of The General Atomic Corporation, and a former director at Los Alamos, published an article in *Science*, denying all claims by Medvedev that the Urals contamination had been caused by the accidental mishandling of P=L=U= T=O=N=I=U=M.[9]

The cover-up by the Nuclear Industry was worldwide. In England, the British Atomic Energy Authority led by its chairman Sir John Hill, said Medvedev's article was "nothing more than rubbish"[10]

In 1979, a team of scientists at Oak Ridge National Laboratory who had been studying the evidence at hand, concluded that there had been an "unmistakable major airborne release (in the Urals) involving moderate to long-lived fission products during the winter of 1957-58."

(It was now 22 years after the event, after all the cover-ups and all the discrediting arguments against the reality of the accident, by the international nuclear business community).

The Oak Ridge team also found, based on the evidence, that "the most likely cause of the airborne contamination was the chemical explosion of high-level radioactive wastes associated with a Soviet military P=L=U=T=O=N=I=U=M site." They found high levels of Strontium-90 and Cesium-137 scattered throughout the

---

[9] Stephen Hilgartner et al., *Nukespeak* (San Francisco: Sierra Club Books, 1982), pp 112–117.

[10] Ties of loyalty among members of the international "nuclear brotherhood" are much stronger to each other, than is their sense of responsibility to the public. The deadly deceit of the British Government in all things nuclear has been well documented by Marilynne Robinson in her book *Mother Country* (New York, 1989).

surrounding area, contaminating water, grasses, air, plants and animals.

The Oak Ridge Team looked at Russian maps of the region before the explosion, and after the 1957 explosion. They found 30 communities had been deleted from the maps after 1957, with a consequent rise in population in the surrounding regions, indicating that large populations had had to be re-located. No other Soviet maps had ever shown this kind of extraordinary movement or deletion.

At Los Alamos in 1978, Medvedev debated Edward Teller. A scientist in the audience rose and warned everyone present that now that Medvedev's information had been released, it would cause great harm to the Nuclear Industry, since its opponents would now use the data against the industry.

The last thing Los Alamos and the Nuclear Industry wanted, was for the facts about P=L=U=T=O=N-=I=U=M to become known.

TWENTY-TWO

ROMANTIC NAMES FOR DEADLY PLACES

P=L=U=T=O=N=I=U=M POISONING is already underway and its effects are being felt world-wide.[11] P=L=U=T=O=N=I=U=M has a special longing for human

---

[11] Richard Miller, in his book *Under the Cloud* (New York and London: The Free Press [Macmillan], 1986) has followed the "Tracks of Selected Trajectories" for both subsurface and atmospheric nuclear detonations that took place between 1951 and 1961. Radiation levels at various sites were measured one hour after detonation.

Given here is only one example out of thousands of such detonation "experiments":

In 1951, Los Alamos sponsored a nuclear bomb test whose trajectory (at 10,000 feet) was studied by the U.S. Government. The test was called "Ranger Able." The area of explosion was Frenchman's Flat in Nevada. The nuclear cloud rose to 17,000 feet, and then the winds aloft took the radiation and spread it around to the following states: Nevada, Utah (Provo), Colorado (Pueblo), Kansas (Wallace), Missouri (very many cities were affected), Illinois tissue. Breathed in, P=L=U=T=O=N=I=U=M produces lung cancer. When it falls on our Earth from a plume of smoke (often, power plant releases may be invisible to the naked eye), or when it runs from underground storage or burial sites, it is eventually soaked up by underground water or rivers (called aqifers) which finally run downhill into rivers and streams, poisoning forever all the drinking water south of itself—if south is downhill.

At first, P=L=U=T=O=N=I=U=M enters into our drinking water in minute quantities which over time add up to ever larger quantities, and in doing so enters our bodies and immediately seeks out human bone marrow and human livers, rendering them cancerous.

(Springfield and Jacksonville), Indiana (Kokomo, Marion and Berne), Ohio (from Springfield to Dayton), Pennsylvania (Williamsport and Wilkes-Barre), New York (Port Jervis, Poughkeepsie), Connecticut (New Haven and New London), Rhode Island (Newport), Massachusetts (Fall River).

A second trajectory from this explosion was tracked at 30,000 feet which followed the first, and fell on the following cities: Caliente, Nevada; Moab, Utah; Colorado Springs, Colorado; Castle Rock and Topeka, Kansas. In Missouri many cities seemed to be affected, among them Kansas City, Sedalia, Columbia and St. Louis. The others were Vandalia, Illinois; Bloomington, Columbus, and Greensburg in Indiana; Marietta and Athens in Ohio; Uniontown, Lancaster and Reading in Pennsylvania; and New Brunswick in New Jersey.

Now, this was just *one* test, out of thousands of such tests.

For an extensive discussion of P=L=U=T=O=N=I=U=M poisoning, let the reader refer to Wasserman and Solomon, *Killing Our Own,* (New York: Delacorte Press, 1982) pp. 177-189.

In addition to nuclear bomb explosions there are, of course, daily, weekly, and monthly leaks (year in and year out) from nuclear plants into the air. There is also the creation of quantities of nuclear garbage from nuclear plants and nuclear defense facilities. These wastes are now so enormous and so overwhelming, that the "disposal" of these wastes has occupied the "thinking" of the Department of Energy for the past 20 years. Their effort to "transport" and then "bury" these wastes where they won't be seen, is understood to create the greatest hazard to the drinking water systems of the U.S.

"Burying" wastes also creates the further hazard of generating volatile gases that may erupt into uncontainable explosions that might one day equal or surpass Chernobyl.

P=L=U=T=O=N=I=U=M is indestructible, alive and kinetic, always moving and broadcasting invisible penetrating radio waves. It has the stubborn and persistent power to alter the healthy electrical balance of our cells. So affected, such cells will forever become crippled, charged and re-programmed by the energy of radioactivity to reproduce—not the normal cells we were given at birth, but now wild and chaotic cells that begin to build in our deepest tissues. It is this wild and undisciplined growth of cells that is called Cancer.[12]

A pregnant woman who breathes in the wake of a P=L=U=T=O=N=I=U=M-poisoned plume of air, or drinks its poisoned water, will transfer that poison to the developing embryo *in utero* which innocently soaks it up. Because an embryo is on the fast-track of cell development, the slightest alteration in the electrical pattern of cell charge will produce embryonic abnormalities. The child is cursed even before emerging from its mother. It will be born either mentally or physically deformed. Its limbs will be crippled; its developing brain structure altered. P=L=U=T=O=N=I=U=M shapes babies in the likeness of its own demented blackness.

---

[12] We know for a certainty that radiation causes cancer. It does so by damaging the DNA coding of cells. DNA is a complex protein stored in the nucleus of each cell. The DNA tells the cell how to function in the body, and also how to reproduce itself. If the DNA is damaged, then the cell loses its program for action and reproduction. The damaged DNA begins to send out distorted messages to the cell, telling it to reproduce mutated clone cells. It is this body of mutants that forms the basis for tumors and a devastated bodily system. By the time a tumor can be detected, it is already composed of millions of mutated cells, all of them obeying the now distorted signals caused by radiation.

If allowed to continue, these cells will eventually overwhelm the body and kill it.

"A pregnant woman who breathes in the wake of a P=L=U=T=O=-N=I=U=M poisoned plume of air, or drinks its poisoned water will transfer that poison to the developing embryo. . . which innocently soaks it up."

P=L=U=T=O=N=I=U=M has an affinity for red-blood cells. That can mean damage or alteration to our immune system. In mature people such damage may bring on slow and painful deterioration and death. Since older people have been accumulating background radiation, radiation from teeth and bone X-rays, and chemical pollution from a variety of sources and adding them all to the body's radioactive bank account (from which no withdrawals are allowed) further doses of P=L=U-=T=O=N=I=U=M-poisoning may tip the scale between toleration and the inability of the body to sustain further insult. In young adults the result is premature aging; in older folks, the wages of the sins of P=L=U=T=O=N-=I=U=M is death.

How safe are we from poisoning? Wherever there is any Nuclear activity, we are in potential danger of spill attack.

It used to be that we were fearful of massive Atomic Attack by our enemies. We are well aware by now of the danger and pain a devastating Nuclear Attack would bring us.

The author (Stanley Berne) was in Japan soon after the Bomb was dropped on Hiroshima. President Truman was advised to bring the war home directly to the Japanese Emperor, the Japanese Army Generals and the people of Japan, in order to stop the slaughter of World War II. The Japanese were told by their leaders that they were winning the war, that it would soon be over, and America would suffer a humiliating defeat. They tirelessly reminded the Japanese people of their great victory over America at Pearl Harbor. The Japanese people were told they were invincible, and that their cause was holy; their mission in life was to rule the world, and to do so was the grave responsibility of the Japanese "race."

Before the Bomb was actually dropped, American soldiers had first to liberate the Philippine Islands from the Japanese. The Japanese were a stubborn, intractable enemy. They were excellent soldiers, unafraid to die, because they were drilled into the belief that an honorable death would send them straight to Japanese "heaven." Kamikaze was in their blood—it meant they would willingly seek death if that would humiliate the enemy—us. Any sacrifice for the Emperor was justified. They were also an unforgivably cruel occupying force in the Philippines. They burned civilian villages, raped women and girls, bayonetted children. It was mindless slaughter. This author interviewed civilians who were driven deep into the mountains where they had to breed and eat dogs to survive, as the Japanese forces occupied the low ground, fortifying it against the inevitable invasion by us.

President Truman had very little choice. It must be The Bomb, a direct hit on a Japanese city, or it meant millions of American casualities, as we prepared to invade the Japanese mainland. While President Truman conferred with the scientists at Los Alamos, we in the Philippines loaded our ships with invasion gear.

After The Bomb was dropped, the Japanese understood their serious miscalculation of America's determination, and they soon surrendered. The American Army immediately turned to the work of occupation, and under General MacArthur we entered Japan as peacefully as tourists on holiday. We could hardly believe our good luck, and we were surprised to find that the Japanese people, weary of the sacrifices of war, bowed and smiled and appeared to actually *welcome* us.

The author, now in Japan and part of the U.S. Army of Occupation, was asked to form a small party that would report on the status of Hiroshima. We got into jeeps and drove to the end of the road which normally would have been the high road into the city. There we were halted by a rope strung across the road on which hung a crude hand-lettered sign in English (and Japanese) which said:

STOP
RADIATION DANGER
DO NOT
ENTER

The sight before us was devastating. There was no city, only a huge blackened and blighted area where once there had been a large city. As far as the eye could see, there wasn't a single blade of grass, or a single bird. No trees stood, only blackened and torn stumps of trees that used to shade the streets. No house stood, no building was left standing. There was only a single brick wall in the far distance that had once been part of the tallest building in town, the newspaper office. The ground before us was baked hard, black and "bald." There was not even a grain of loose earth or gravel. All had been swept clean by the powerful blast.

There was complete silence, not a single sign of life anywhere. The air and the Earth had been turned a surreal grey. It was a landscape of nothingness, slightly overhung with light mist or smoke.

We have had our minds trained on the greatest *immediate* danger of the 20th century— direct Nuclear Attack. But this has slowly changed with the advent of applying Nuclear Energy to everyday "peaceful purposes." Now, perhaps, the greatest danger of all may well be the small and carefully ignored Nuclear Attacks that insult us every day. Every release from a Nuclear Power Plant, is an Atomic Attack; every barrel of trans-uranic waste generated daily by the Utilities, is a Nuclear Attack; every large steel drum full of high-level nuclear waste, is an Atomic Attack; every thousand gallons of liquid lubricant and water used to cool atomic cores, is an Atomic Attack; every cardboard box stuffed with contaminated machine parts and discarded nuts and bolts from the plant, is a Nuclear Attack; all the uranium tailings thrown up alongside the old mine, and the abandoned enrichment factory, left to leach out in the rain in summer, and the snow in winter, is an Atomic Attack; the urine colored streams that seep out of Rocky Flats or any P=L=U=T=O=N=I=U=M smelter or refinery, is an Atomic Attack; every $50,000,000 incinerator thoughtfully and even delicately designed at Los Alamos to *burn* P=L=U=T=O=N=I=U=M waste, because its sheer bulk is such an embarrassment to the Utilities, thus effectively storing P=L=U=T=O=N=I=U=M in the sky, is an Atomic Attack—and the DOE plans to build and sell these incinerators to every state in the union, in order to bypass the distaste people have to transporting Nuclear waste past their businesses, their homes, schools and churches. Burning P=L=U=T=O=N=I=U=M in the sky could add up to and equal an above-ground Nuclear Bomb explosion.

These everyday spills are not the big-bang single-shot Nuclear Attacks that will end the world—these are little bangs, the tiny additive insults that we are expected to share our air with for the convenience of Nuclear Power. We are being asked, each day, to co-exist with P=L=U=T=O=N=I=U=M. We are asked to go as far in toleration as our bodies will bear—and no one knows how far our bodies can be stressed before our cells are simply overwhelmed.

The U.S Council for Energy Awareness, the trade association of the Nuclear Industry, shows a total of 112 operating "Units" or Nuclear Power Plants. They claim to be producing as much as 19.7% of the total U.S. electricity generated. They never mention down-time. There is maintenance down-time, repairs down-time, emergency down-time. If down-time is figured in, they are producing, perhaps, 11% of the total electrical energy in the U. S. today.

Their trade association map shows the location of every Nuclear facility in America. Very few states are left out. Those states that are left out are the least populated ones. The more population, the more Reactor Plants, and inevitably, the more Nuclear Waste and the more P=L=U=T=O=N=I=U=M is generated. Montana is exempt, Wyoming, Utah, Nevada, New Mexico, North and South Dakota are exempt. And a few other states, very few, are exempt. Where the population is very dense, there sit the greatest number of plants.

> California has 6. Illinois has 15. Connecticut has 4. Florida has 5. Massachusetts has 2. Michigan has 5. New Jersey has 4. New York has 7. Pennsylvania has 9. Alabama has 7. Arizona has 3. Georgia has 4. Minnesota has 3. Nebraska has 2. North Carolina has 5. Ohio has 3. South Carolina has 7. Tennessee has 4. Virginia has 4. Washington has 4. Wisconsin has 4. How many are there in your state?[13]

It is the conscious policy of the industry to give their plants very cute and cuddly names, even romantic ones. These are benign names for a growling, threatening, restive beast on a skinny chain: Browns [sic] Ferry

---

[13] There are *no* nuclear power plants in Nevada, Utah, Montana, New Mexico, North Dakota, South Dakota, Oklahoma and West Virginia. However, some of these states are suffering from severe problems brought on by the accumulation of nuclear waste, transported through these states, or deposited in these states, by the Department of Energy.

For example, Idaho has a severe contamination problem because it is being used as a dump site by the DOE. Nevada has been struggling for years with the DOE to try to prevent their being turned into a national dump site. Additionally, Nevada has had to deal with the nuclear bomb testing sites established by the DOE over the protests of its citizens.

in Alabama—Palo Verde in Arizona—Diablo Canyon in California (no doubt, the Devil's own headquarters)—Fort St. Vrain in Colorado—Millstone in Connecticut—Turkey Point in Florida City—Zion in Illinois—Wolf Creek in Kansas—Maine Yankee in Maine—Prairie Island in Minnesota—Oyster Creek in New Jersey—Trojan Nuclear in Oregon—and our favorite: Peach Bottom Atomic Power Station, Units 2 and 3 in Pennsylvania.

Don't these sound like fun places? They are names more suitable perhaps to ski resorts, or fishing villages—places where Americans might go to relax and have fun.

Alas, they are not fun places. They are places of deep morbidity, centers of immense potential danger. Places where, should anything ever go wrong, the dread result would be widespread death and disease.

TWENTY-THREE

FROM THE ARCTIC TO THE GANGES RIVER —AS BLOWS THE WIND

ONE OF THE MOST POWERFUL trade organizations of the Nuclear Power Industry is the Atomic Industrial Forum (AIF) which has, as its mission, the "education" of the general public.[14] It takes its work

---

[14] The AIF has undergone a name change recently. It now calls itself, the U.S. Council for Energy Awareness. Their address is Suite 400, 1776 I Street, NW, Washington, D.C. 20006-2495.

While its name may be changed, its mission remains the same: to present the good news on all things nuclear. Their elaborate brochures and their extensive media advertising call Nuclear Power the answer to Arab oil. Both conservation and alternative energy programs are condemned as "band-aids." The Council declares nuclear power to be the cheapest (!) and cleanest (!) method of preserving America's "way of life."

seriously, has a very large budget, but is very secretive about the amount. Its goal is to persuade the American people that Nuclear Power is inevitable: so relax and enjoy it.

The "language" adopted by this arm of the industry is both awkward and unfortunate, since it seems to employ double-speak almost exclusively. According to this agency, there are never any Nuclear "accidents." Instead, there may be the occasional "incident." If anything serious ever happens, this agency is called upon to limit the damage with words, by inventing benign phrases to cover over reality. For example, the AIF never has admitted there was a serious hydrogen explosion at Three Mile Island. They called it "An Energetic Disassembly." There was no fire either, but they did admit to an "Episode of Rapid Oxidation." When the fuel assembly, the heart of the reactor core, became virulently contaminated with radioactivity as did the surrounding reactor vessel: not to worry. They announced that as far as the reactor was concerned, what happened was that P=L=U=T=O=N=I=U=M "Took Up Residence" in the vessel.

For the AIF, an explosion is a "power excursion." If it is serious and causes costly destruction of the plant facility, it becomes a "Disruptive Energy Release."

The AIF also has the mission to disparage and diffuse all critics of Nuclear Power. The AIF has characterized its critics as "idle, fearful, timid, and against the American Way of Life."

The AIF has millions of dollars for advertising the advantages of Nuclear Power. This is a serious concern, since it is generally how the American public (and the press) gets its knowledge of the Nuclear business. The information is, of course, steeply slanted in the direction of the interests of the Utilities. Accordingly, Nuclear Power is always cleaner, safer, better than any other form of generation.

The AIF never discusses the question of "radiation," and is at pains to avoid mention of the immense hidden costs supporting Nuclear Power which are paid for by the taxpayer. There is the very heavy cost of research and development (congress regularly appropriates billions of dollars for this under the aegis of national defense) conducted at government laboratories like Los Alamos, or Sandia National Laboratories at Albuquerque, New Mexico; the expenses of uranium enrichment and even its importation now from the Soviet Union; the reprocessing of high-level and low-level waste, and its transportation from plants all over the country to some supposed central depot for underground storage; and there are the tax advantages Utilities enjoy, which must be made up by other businesses, as well as by individual taxpayers.

The AIF coordinates with the public relations staff at every Nuclear Power Plant. The public is thus offered a continuous dose of Pro-Nuclear "propaganda." It is rightly propaganda, since it is designed far less to inform than to persuade.

The AIF has always said that the P=L=U-=T=O=N=I=U=M "that doesn't reach you is completely harmless." Well, the same goes for arsenic. It doesn't poison you until you swallow it. But, arsenic is safer than P=L=U=T=O=N=I=U=M, since your senses warn you of its presence. Not so with P=L=U=T=O=N=I=U=M. Once created, P=L=U=T=O=N=I=U=M remains alive and volatile *forever,* emitting alpha particles that give no hint or warning of its presence. It is odorless and invisible. You might suddenly be dosed, and never know it until years later, when the terrible pain begins.[15]

The repeated argument of the AIF is that background radiation, existent in all of nature, is responsible for the reported increase in cancer and genetic damage, not P=L=U=T=O=N=I=U=M. Of course, what they neglect to tell us is that when you raise the dose, and then add it to existent background radiation and X-rays, cancer inevitably results.

The Industry provides their own Pro-Nuclear trade commissions and associations with all the money they can use to fight and overcome the almost penniless citizen opposition groups. A single Pro-Nuclear public relations agency, the Edison Electric Institute was known to be budgeted at $14.6 million in 1979. Citizen information and research groups operate on the few thousand dollars they manage to scrape together from ordinary citizens.

The Utility Corporations well know their nuclear plants are lethal, along with everything in it, and they have tried to think of everything that might contain fire, explosion and radioactive floods. And yet, there appears to be a mean spirit coloring the activities of the Nuclear

---

[15] Of course, P=L=U=T=O=N=I=U=M is a metal, heavy and grey in appearance. It is this metal which emits invisible deadly activity. The industry itself goes to ingenious lengths not to come into contact with it. However, P=L=U-=T=O=N=I=U=M has crafty ways of slipping past all precaution.

Industry. Deception, secrecy and tortured information are apparent constants in their public positions and announcements. This has to be viewed as a serious attempt by the industry to hide their deficiencies, even as public health is central to this issue.

"Deception, secrecy and tortured information are apparent constants in their public positions and announcements. This has to be viewed as a serious attempt by the industry to hide their deficiencies. . ."

When three senior General Electric engineers quit their jobs in February, 1976, they said:

> The cumulative effect of all defects and deficiencies in the design, construction, and operation of nuclear power plants makes a nuclear power plant accident, in our opinion, a certain event.

One of the engineers, Dale Bridenbaugh said, he had grown "increasingly alarmed. . . . Nuclear Power [had become] a technological monster."

What the engineers were alarmed at, were the defects in General Electric's Boiling Water Reactors, such as:

—pipes that cracked
—seals that leaked
—valves that stuck and stayed open
—unpredictable vibrations that shook parts loose from their fittings and joints
—control rods that were installed *upside down*, rendering them useless to control a meltdown or explosion
—defective welding
—fuel rods cracking and shrinking
—loose switches
—faulty wiring

> The belief that the public had been denied information about Nuclear Power was quickly transformed into an assumption that the nuclear community had something to hide. The arrogance of the nuclear community added to this belief; in every industrial nation, the nuclear communities regarded themselves as a special technical elite.
>
> *The Nuclear Barons*
> Peter Pringle and James Spigelman

In every case of radioactive release into the environment—*in every single case*—information about the release was guarded from the public, and smothered in official secrecy. This seems an intolerable situation, since our personal health and safety is at risk, and yet we are kept from any knowledge of that risk.

The spills have been many and they cross national boundaries, as do the air currents that carry them aloft. There were secret releases at the following plants:

—Dresden Nuclear Station, near Chicago
—Turkey Point in Florida
—The Millstone Plant in Connecticut
—The Wurgassen Plant and The Brunsbuttel Plant—both in Germany
—Plants at Tsuruga and at Fukushima in Japan.

There were further environmental failures at fuel reprocessing centers at:

—West Valley, New York
—Barnwell, South Carolina
—Karlsruhre, Germany
—The Tokai Plant in Japan
—La Hague Plant in Cherbourg, France.

There was also the near fatal fire at Rocky Flats P=L=U=T=O=N=I=U=M bomb and smelter center in Colorado in 1969, which caused widespread P=L=U=T=O=N=I=U=M poisoning for miles around the countryside, wherever the wind blew the poison.

Uranium tailings polluted all the drinking water at Rum Jungle, Australia, as well as at the Manhattan Project Uranium Processing Plant, at Port Hope, Ontario, Canada. That spill poisoned beaches, whole towns, homes and schools for miles around.

There were technological failures in unlikely places, where none were ever expected. Soviet "experts" were confounded when in 1978 their Satellite, Cosmos 984, exploded in the skies over northern Canada, spewing out radioactive particles over the Arctic. The CIA was equally taken aback when their "foolproof" Nuclear Powered Spy Station, focused on China, exploded, polluting the Ganges River in India.

TWENTY-FOUR

A LITTLE $35 BILLION DOLLAR ERROR

THERE WAS GREAT HOPE after World War II that we were entering a new era of peace and plenty since new uses were projected for atomic power. The atom had won the war, and a grateful people were optimistic that the atom would win the peace. It was to be an era of greater prosperity for all, and a time when mankind could look forward to less toil and more leisure. It was to be an era of cheap power.

Power has always been conceived of as issuing from a central utility source that would produce and distribute it, at (hopefully) cheap rates to the consumer, and with great profits falling to the giant utilities, their shareholders, and (especially) to their directors.

A frenzy of building atomic power plants got under way in the late seventies and eighties that was thought would usher in the new era of prosperity.

Aboard *Air Force One*, Howard Baker, Chief of Staff to President Ronald Wilson Reagan delivered the comics to his President during a flight in 1988. It is this President, and this administration, which signalled to the utility companies of America that it was all right for them to build Atomic Power Plants "their way."

Utilities have always wanted to build their plants "their way," which meant, build them as near to their customers as they could get. This makes good utility sense, since there is a saving to the company on shorter wiring, and shorter wiring means an even greater saving on time and maintenance.

In the 1980's utility executives were planning Nuclear Power Stations in the old way, as if they were the oil-fired power stations of the past. They began with the premise that saving money was more important to the corporation than anything else, even safety. Besides, optimism about Nuclear Power was so contagious and so blinding, that they carried on in the tradition of Glenn T. Seaborg, the first chairman of the AEC who "knew" that Atomic Power was "inherently" safe and really "invulnerable" to accident. They were supremely confident that their form of engineering was going to leave nothing to chance, that they could make these plants models of both profit and safety.

With this philosophy in hand, the Long Island Lighting Company of New York, began to build a Nuclear

Plant on property they owned that lay in the very heart of one of the most heavily impacted population centers in the nation—Long Island—an area very close to the heart of New York City.

The area they had chosen gave little access for escape should there be a need for evacuation of the Island. There is a main highway, State 495, which leads to New York City on one end, and dead-ends near a town called Riverhead, which is on Great Peconic Bay. Apparently, no serious thought was ever given by the Utility executives to a possible failure at the plant. If a disaster were to occur, then people would be caught in the historic gridlock traffic jam of the century. As P=L=U=T=O=N=I=U=M rained down on the area from an explosion (the worst case scenario) then Long Island would have to be abandoned forever, as happened in the Ural Mountains of Russia. There, people were forced to leave their homes, their farms, their machinery, their animals, their belongings, and had to be trucked out of the area with only the clothes on their backs. Then, the area was burned down. With the population density of New York, where would the people of Long Island go? Would they fight and wrestle their way into New York City to wander the streets, so many millions homeless? What terrible disasters would flow from that possibility?

Nothing daunted, the thoughtful executives of Long Island Lighting went ahead and began to assemble the plant. They *finished* it in 1988 at a cost of $5.3 billion.

It dawned on the people of Long Island that they had the makings of a monumental disaster in their midst. That disaster, which presses on the heart of all nuclear power, raises the question of tolerating a loathsome method of generation at the center of a heavily impacted civilized community. Is that a good solution to the power needs of any community? The people of Long Island, the State of New York, and Governor Cuomo of New York all said NO! It was the citizens of Long Island who finally evaluated what had been going on, and who rejected the opening of the plant. They simply demanded that the completed plant be shut down and removed.

This "incident" doesn't exactly allow us to believe that Nuclear executives and the engineers they employ to assemble power plants, have all their oars in the water when it comes to the planning and operation of what might well be a lethal power factory in the backyard of millions of people.

Long Island Power built the plant, spent $5.3 billion doing so, and then were refused a permit by the State of New York to operate it. This was the first completed plant in the U. S. to be shut down just before coming on-line. It was obviously a gross administrative failure. The plant and all its equipment was sold to the State of New York for one dollar. Then, the people of Long Island paid for that mistake with a 5% hike in their utility rates. With that hike, the Utility Company plans to recoup more money than it wasted on the plant.

This case was, alas, far from unique. Careful planning does seem to be in rather short supply in the Nuclear Business. Two dozen other plants were closed down during construction, mostly due to management error. Altogether, Utilities have lost over $35 billion on failed plants.

In South Carolina, the Duke Power Company spent and lost $700 million on their Cherokee Plant near Gaffney.

In Michigan, in 1969, Consumers Power Company began to build the Midland Nuclear Plant near Lake Huron and was still trying to get it started ten years later, after having spent more than $4.1 billion. The plant was coming in twelve times over budget. They finally decided to quit trying to start it and converted the plant to burn either coal or oil.

In Tennessee, the TVA spent more than $2.4 billion trying to assemble its proposed four reactor plant at Hartville, 40 miles from Nashville. They couldn't get it going, and so abandoned it. They had no money left over to raze the structures they had built because it is very expensive to tear down cooling towers made of heavily reinforced concrete. It's cheaper to let them stand and rot.

In Indiana, the Marble Hill plant was finally abandoned, and Washington State abandoned its Nuclear Plant, after actually getting it started. They stopped just as they were warned that the plant was incorrectly assembled, and was poised for a disaster.

The Palo Verde Nuclear Generating Station is the most expensive Nuclear Power Plant in the nation. It cost $10 billion. It is located 50 miles west of Phoenix, Arizona. It may be a disaster waiting to happen. The Nuclear Regulatory Commission, which is markedly friendly to Utility Corporations, found numerous safety violations at the plant in July, 1990:

—The fire protection plan actually prevented a shut-down in emergencies
—15 areas in the plant are obstructed and cannot be reached during a fire
—Workers are expected to manually operate automatic switches and valves during an emergency
—To shut down the three reactors, workers would have to wade through fire and smoke
—There are major weaknesses in areas of maintenance and surveillance, technical support and safety assessment
—There is poor quality verification
—The lighting of the complex is faulty
—Emergency batteries for lighting were found to be dead
—Emergency lighting failed to come on.

Palo Verde was financed by a consortium of utilities: Arizona Public Service, El Paso Electric, Public Service of New Mexico, The Salt River Project of Arizona, Southern California Edison, The Los Angeles Department of Water and Power, and The Southern California Public Power Authority. So much money has been lost by these corporations that they fought to recover their losses by passing them on to their respective rate-payers in their areas. Consumer groups fought back, and some of these companies went to the brink of bankruptcy.[16]

---

[16] In one particular case, the city government of Albuquerque, which is the largest customer for electricity in the State of New Mexico, objected when Public Service of New Mexico proposed to raise its rates in order to pay for its losses on Palo Verde. The New Mexico Public Service commission ruled that the investment in Palo Verde was an administrative error

It is certainly sad that as the 20th century comes to a close, the fabled plenty of "The Nuclear Age" has seemed to come to a sorry end. Nothing daunted, however, the Department of Energy is still briskly carrying on its mission of encouraging the Nuclear Industry to produce P=L=U=T=O=N=I=U=M in ever larger quantities to enable it to make and test "a new generation" of Atom and Hydrogen Bombs.

The world is, indeed, a dangerous place, and our country needs to be prepared to defend its interests in oil and general business and trade. But that very danger must include the death-threat inherent in the continued production of P=L=U=T=O=N=I=U=M, anywhere in the world.

---

on the part of Public Service, and the rate increase was rejected. As a result, Public Service of New Mexico came close to declaring bankruptcy.

So far as citizen groups are concerned, it is their action alone that has slowed down the growth of the nuclear industry. Citizens for Alternatives to Radioactive Dumping (CARD), in Albuquerque, Concerned Citizens for Nuclear Safety (CCNS) in Santa fe, New Mexico, Citizen Alert in Reno, Nevada, Southwest Research, another citizen supported group in Albuquerque (to name but a very few of the many active citizen groups)—all of them all over the U.S. have succeeded in questioning the motives and safety of transporting P=L=U=T=O=N=I=U=M-infected waste across the nation, and burying it in Nevada and New Mexico.

The industry claims it must get rid of the waste, to make room for more waste. Citizen groups have said, the only safe thing to do with P=L=U-=T=O= N=I=U=M is *not make more.*

TWENTY-FIVE

WILD SCIENCE, ROGUE TECHNOLOGY

FOUR SERIOUS TECHNOLOGICAL FAILURES have given the world much to weigh and ponder about the freedom of engineering to apply its own discreet solutions to human problems without independent public oversight. The failures are clearly Chernobyl, Three Mile Island, Love Canal and Bhopal. These technological installations went wildly out of control, and the resulting disasters were powerful enough in themselves to reshape the debate over energy development, and environmental policy.

The event that concerns us here, is the catastrophe at the nuclear power plant at Chernobyl, in the Soviet Union. On April 26, 1986 there was a meltdown, fire and an enormous explosion in Reactor No. 4, an event which Russian nuclear technocrats "guaranteed" would never happen.

Chernobyl is without parallel as a technological disaster in the modern world. 300 persons died immediately, and 100 million Soviet citizens were exposed to radiation and its characteristic immediate and delayed effects. 28,000 people have already died of delayed cancer fatalities. 116,000 people have had to be evacuated from the surrounding areas in the Russian Ukraine, and communities as far away as 2000 kilometers have had their food and agriculture destroyed forever.

"Chernobyl is without parallel as a technological disaster in the modern world."

The direct and indirect costs of the disaster continue to mount, and it is possible that they will clearly exceed *the total Soviet investment since 1954, in creating its nuclear capacity*. Not only that, but the effects of radiation poisoning were discovered in communities throughout Central, Western and Northern Europe. Radiation "rain" has been measured in the Cumbrian Mountains of Northern England and in the dairy pastures of West Germany. In Northern Sweden the Saami people found that 275,000 of their reindeer (on which their economy entirely depends) had to be slaughtered and disposed of because the lichen on which they fed was contaminated with Cesium-137.

Nine days after the explosion, Chernobyl fallout reached the United States. The highest amount of radioactivity was discovered in rainwater at Spokane, Washington, where readings rose to an alarming 6,600 picocuries per liter by May 12, 1986.[17]

Chernobyl "rain" continued to fall throughout the U.S. By May 16th low-level radiation was recorded

---

[17] By the mid-1930's it became clear that radiation was a real threat to humans and that measurements were necessary. To express the limits to which people working with radiation could be exposed, scientists developed curies, roentgen units, rads and rems.

The curie (Ci) is the amount of radioactivity in a gram of radium, which degrades at 27 billion disintegrations a second. There are billions and billions of curies in an atomic reactor.

The curie is often broken down into smaller units, with one curie equal to 1000 millicuries (mCi), one million microcuries (uCi), or one trillion pico-curies (pCi). The curie measures gross radiation, not biological damage.

The "roentgen" measures the amount of radiation affecting a "target" struck by radiation.

The "rad" stands for *radiation absorbed dose.* Another abbreviation "rem," stands for *roentgen equivalent in man.* Rems, rads and roentgen units are very similar and are used to measure biological damage from radiation, with perhaps the *rem* being used the most, to measure biological damage. *A rem equals 1000 millirems.*

Because the units may be too large for certain uses, the prefix milli (m) is often used with roentgens, rads and rems to indicate *smaller* quantities. One rad or rem equals 1000 millirads (m rad), or 1000 millirems (m rems).

by 50 Environmental Protection Agency milk monitoring stations.

The nuclear industry in the United States has always dismissed the effects of low-level radiation from both fallout and from nuclear plant releases. Low-level radiation simply means, that populations are exposed to incremental radiation doses that over time, can add up to and even equal the single dramatic explosive event of a nuclear bomb attack.

The event at Chernobyl occurred in 1986, and the chronology went like this:

April 25, 1986.

> For some unknown reason, perhaps for purposes of testing, the cooling system that kept the reactor in control, was turned off.

The Nuclear Regulatory Commission has just issued new radiation-exposure limits for both nuclear plant workers and residents in the area of the plants. These regulations are supposed to go into effect by January, 1993. (They were announced on December 13, 1990.)

According to U.S. Government studies, the average American can expect to be exposed to more than 360 millirems of radiation a year, from various sources. (Radiation from X-rays account for about 6 to 7 millirems a year.)

The new rules are supposed to set a limit of 100 millirems a year on human populations. (They are now exposed to 500 millirems a year.)

Nuclear workers are now exposed to 20 rems (20,000 millirems) of radiation per year. The new rules advocate 5 rems.

Michael Mariotte of the Nuclear Information and Resource Service claims that the NRC is actually allowing *higher* amounts of radiation to be released into the air, by not controlling the *range* of higher concentrations of radioisotopes released into the air, which is allowed, under the new rules.

Critics of the NRC charge the agency with enacting rules made 13 years ago by the International Commission on Radiological Protection (1977).

Plant operators are not at all *required* to follow the new rulings; they are only being urged to do so. *There is no present law requiring plant operators to limit nuclear emissions.*

Not only that, but the technicians apparently *forgot* to turn the wheel that would bring the cooling system back on. As a result of human error, the reactor was kept working for eleven hours, growing ever hotter. Either the temperature gauges failed to indicate that heat was building up in and around the reactor, or no one bothered to read the thermometers.

April 26, 1986.

The reactor, now untethered and on its own, lacking all human constraints, did what any reactor would do in this circumstance: it blew! At 1:23 in the morning, reactor no. 4 exploded, producing a hydrogen fire so hot and so devastating, that the plant authorities simply did not know how to control it. They informed the Communist government officials of their problem, and the officials responded by urging silence. Even though the Government knew that levels of radiation were rising and spreading—they kept completely silent.

April 28, 1986.

At monitoring stations in Sweden, Denmark and Finland it was noticed that suddenly there appeared to be unusually high readings of radioactive levels. The Soviet government finally announced that a nuclear "accident" had taken place, but no mention was made of radioactivity.

April 29, 1986.

The Soviet government concedes that there is a fire of enormous proportion at Chernobyl Power Station in the Ukraine, in the province of Byelorussia, and that some "leakage of radioactive substances" has taken place. They also admit that they don't know how to quench the fire, and request immediate international assistance in atomic fire control.

April 30, 1986.

No further information about the event at Chernobyl has been disclosed. There is still an attempt by the Soviets to cover up the accident. However, Western European foreign and defense ministries demand that the Soviets be forthcoming about the disaster, and offer details, so they can take defensive measures to protect their own nationals.

May 1, 1986

Soviet helicopters pour wet sand over the fires.

May 6, 1986.

Soviet officials concede that earlier reports were clearly understated, and begin to publish detailed reports of the accident.

May 7, 1986.

Now that the European community is aware of what happened, they take measures to protect their own populations as best they can. They ban all fresh produce from Russia and from surrounding areas and countries.

May 15, 1986.

President Gorbachev goes on national TV, 18 days after the event, to inform the nation of the extent of the disaster.

July 20, 1986.

The Russian Politburo begins criminal proceedings against the management at Chernobyl.

August 26-28, 1986.

The International Atomic Energy Agency meets in Vienna, and Western scientists begin to piece together the possible chronology of the accident: they determine that there was a steam explosion in the boilers, which led to a hydrogen gas leak and explosion, and that the chief cause of the accident was human error. They accuse the Soviet Government of sacrificing safety to energy production.

July 29, 1987 (one year later).

> The head of the Chernobyl nuclear power station and two aids are found guilty by the courts and sentenced to 10 years each in a labor camp. It was revealed at the trial that the management failed to evacuate the workers, and that as a result many died.

April 26, 1987.

> The Supreme Soviet votes $26 billion (dollar amounts) for clean-up, evacuation of refugees, and the building of health clinics and hospitals, over the next five years.

June 21, 1990.

> The Soviets agree to set up an international study center to follow up on the long-term effects of radiation and its diseases.

If you could visit Chernobyl today you would find Reactor No. 4 has been entombed in a towering bulk of cracked reinforced concrete that rises up 200 feet above the sandy soil. Covering the concrete is an ugly skin of rusting steel plates. They are trying to contain the radioactivity in the bulk, but they are not successful. It will take more concrete and still more steel plates, and still they are not sure that that will do the job. The reactor core contains more than 455 pounds of P=L=U=T=O=N-=I=U=M. It will be hundreds of centuries before the P=L=U=T=O=N=I=U=M dies down, and each further entombment will cost the government one billion rubles.

Chernobyl was an old town along the broad Dnieper River. It was a one-time summer resort, where picnickers and campers would come from Kiev, 75 miles to the South.

The disaster has brought total change to the area. So far, the government has spent $15.4 billion to clean up the site. Medical problems have grown beyond the ability of the government to handle the volume—$26 billion has been spent so far on health related matters. The local government estimates the damage at 82 billion rubles, and it needs another 17 billion rubles to re-house displaced persons.

12.3 million acres of land in the Ukraine are no longer habitable. 3.5 million hectares of rich farming land has had to be abandoned, perhaps forever. Byelorussia (the province) has lost 1/5 of its arable land, and according to Russian scientists, 15% of its forests will be contaminated for at least the next 100 years. 27 towns and 2,697 villages are dangerously contaminated. Fish are completely gone from the rivers, and the wild fruits and berries that are able to grow, are radioactive. Tens of thousands of cattle and sheep have had to be destroyed. Dr. Yuri Spizhenko, the Ukranian health minister says 4,000,000 people are still living in affected areas, and estimates 50,000 additional cases of fatal cancer to come.

The distressing statistics continue: 150,000 people have lost their homes; 250,000 people are awaiting resettlement; the Byelorussian Parliament has declared its republic an ecological disaster zone and has appealed for international assistance.

raise, touch, breathe and eat is radioactive. It is common to see eight legged horses, and headless cattle.

The questions raised by Chernobyl concern how any sane individuals could ever argue in favor of nuclear power.

★★★★★

There is here a related matter that begs the question: is Science too important, too personal to all of us, to leave to the scientists?

Aside from Nuclear Power which grew secretly at the will of technocracy, with no oversight or independent surveillance from the people who might someday be directly affected by such growth, there is also the matter of the development of a kind of wild science, a rogue technology that has its own ends, its own goals, its own advocates, its own funding and is permitted to self-create, apart from the possible destructive effects on the health and well-being of the general population, that has no veto power over such growth.

A perfect example of "wild science" in the United States, occurred in 1957 at the Rand Corporation at Santa Monica, California. There, in a cozy sort of collegiate setting and atmosphere right on the beach, a physicist named Nicholas C. Christofilos from the government laboratory at Livermore conceived the notion of exploding 1000 atomic bombs a year into the earth's magnetosphere (just above the atmosphere). His idea was to create a shield of electronic particles which would circle the earth, and according to his reasoning, would serve to protect us from incoming enemy atomic bombs. Dr. Christofilos never mentioned fallout, or the effects his explosions of atomic bombs would have on us who have to live beneath his canopy of particles.

"Dr. Christofilos never mentioned fallout, or the effects his explosions of atomic bombs would have on us who have to live beneath his canopy of particles."

One would think that at least one sane scientist at the Rand Corporation would have rejected the theory as too devastating to human populations, including our own, whom Christofilos was trying to save from very much the same kind of bombs. Alas, in 1957, there was apparently not even one sane scientist at the Rand Corporation to object to the obvious effects such a system would have on life on earth.

None of this would matter very much if the Rand Corporation was some ineffective science department at some backwater university. Alas, again. The Rand Corporation was fed by the Pentagon's "Advanced Research

Projects Agency." The "thinking" that went on at the Rand Corporation was translated into action by the Pentagon, and so they decided to act on Christofilos' interesting theory, and made immediate plans to do so.

Under the rubric "Argus" the scientists set sail for the South Atlantic to test the theory. Now, the Hiroshima bomb was equal to 13,000 tons of TNT. The Rand Corporation people wanted to be conservative. They decided to explode a series of "small" atomic bombs equal to only 1000 tons each of TNT into the upper atmosphere, at an altitude of 200–500 kilometers above the sea. The electrons released by the bombs did indeed form artificial Van Allen Belts above the earth, that were between 90–150 kilometers thick.

The Rand scientists really enjoyed the results of Argus, and wanted to go on to bigger things. They decided that since Argus proved to be such a whopping success, they would enlarge upon it and launch a new series of bombs, under a project which they now named "Starfish."

Starfish was executed on July 9, 1962 above Johnston Island in the Pacific. In this test they exploded *a hydrogen bomb equal to one and one half million tons of TNT into the upper atmosphere.* (Remember Hiroshima? That bomb was only 13,000 tons.)

The effects of Starfish on the upper atmosphere lasted for many more years than this group of scientists were able to measure. In fact, they have never been able to determine how long the pollution belts they created did last. What they did, however, was to launch artificial radiation belts that circled the earth for an indefinite period of time.

Some years later, a scientist at Los Alamos named Herman Hoerlin was asked to write up a secret report on all these matters. He did so. The report was "classified" for many years and finally was "de-classified." In the report, the author estimates that as a result of these tests, space was seriously contaminated. He concluded that astronauts would suffer radiation as they passed through the belts. Not only that, but they would absorb at least 50 rads of radiation *per orbit*. (The *yearly* maximum in space is said to be 27 rads *per year!)* Dr. Hoerlin concluded that Starfish was "a test out of control."

There's more.

In the summer of 1958 the General Atomic Company, which builds nuclear reactors, decided to build a spaceship prototype that would be powered by nuclear bombs. The project was headed by a respected physicist named Freeman Dyson.

The Pentagon thought highly of this project as well, and funded it through its Advanced Research Projects Agency. The company hoped to build a spaceship that would carry "freight" to the moons of Jupiter, and sail on to the star Alpha Centauri—at the per-pound cost of freight that would be equal to the cost of airmail to Europe!

Fortunately, the nuclear-test-ban treaty put an end to the project. However, France continued to explode Nuclear bombs into the upper atmosphere well into the 1970's, and the Chinese did not stop exploding nuclear bombs in space until 1980.

★★★★★

The results of all this "science" gone awry, were Chernobyl, nuclear bombs exploding in the atmosphere above us, and the notion of nuclear-bomb-fueled space vehicles. But, as usual, innocent citizens of the world, by remaining innocent, and by allowing themselves the luxury of ignorance, paid the price. Some investigators have carried the known effects of P=L=U=T=O=N=I=U=M poisoning as far as speculation could carry them. Two such scholars, Dr. Jay Gould and Benjamin Goldman in their book *Low-Level Radiation, High Level Coverup* (New York, 1990) link the accumulation of low-level radioactive fallout from the 1950's, to today's variety of immune deficiency diseases.

We now know that cancers develop long after radiation exposure events. Such problems as breast cancer and thyroid cancers are in this class. *But because these types of cancers can take up to 30 years to develop,* not enough time has passed as yet for such malignancies to be linked directly to P=L=U=T=O=N=I=U=M.

Gould and Goldman, looking at AIDS, for example, as a disease new in this century, having been identified only since 1960, link it to radiation fall-out. They find that nuclear power plus the weapons programs have assaulted human immune systems world-wide. Another clue, for them, is the rise in related immune deficiency diseases such as Chronic Epstein-Barr Virus, Lyme Disease, Herpes and Septicemia. The Soviet Government is now reporting larger numbers of deaths from Pneumonia and Leukemia all over the Soviet Union, since Chernobyl. These are immune deficiency diseases.

The two investigators discovered a rise in Leukemia in 1970 in the Albany-Troy, New York, area, where elevated levels of radiation were measured and linked to the tests conducted as far away as the Nevada Atom Bomb Test Site. They also found that large radiation releases from the Millstone Nuclear Reactor in Connecticut, which had been kept secret, led to elevated cancer mortalities in the nearby townships. They also cite a 300% rise in childhood cancers after 1955 due to the explosions of nuclear bombs in the upper atmosphere. They also show that after Chernobyl, in May of 1986, the United States had the highest annual increase in deaths in 50 years.

★★★★★

Whatever statistics show or cannot show, prove or cannot prove, it must appear that what unregulated Science and profit-driven Technological Engineering are doing to change the earth, must be of deep concern to us all: to ourselves, to our children, and to our world.

Science and Engineering, most often pursued for profit, and directed by businessmen who can easily be persuaded to set aside social responsibility for their actions in favor of dollar amounts, should have to submit their implementations to a combination of citizen and independent scientific oversight, with enough power to veto what might prove painful to all living things that depend on the earth for air, food, water, shelter and survival.

The men who create power make an indispensable contribution to the nation's greatness. But the men who question power make a contribution just as indispensable—for they determine whether we use power, or power uses us.

—President John F. Kennedy

Ours is a world of nuclear giants, and ethical infants.

—General Omar Bradley

A NOTE

SELLAFIELD IN NUCLEAR ENGLAND ENJOYS "CROWN IMMUNITY"

IN 1957, IN ENGLAND, a very dirty nuclear plant named Windscale had succeeded in developing a reputation for disaster. It was home to over 300 nuclear accidents, including a very serious core fire, the most serious before Chernobyl. The fire went out of control for two days, contaminating England, Ireland and Western Europe. British subjects were told *nothing* of the disaster. It was all kept secret. Later, at least 260 cases of thyroid cancer were reported and attributed to the explosion. Those were the *known* consequences.

The Government of Britain decided after that to rename the plant *Sellafield* and to very greatly expand its activities. It has now become Great Britain's *National Atomic Energy Center*, the largest in all Europe. It is located just off the Irish Sea, on the British coast, and lies in the very heart of England's beautiful "Lake District." This geographic area is a National Historic Park, made "immortal" by England's greatest poets, Wordsworth, Coleridge, Tennyson, Keats, Shelley, and its prose writers, Carlyle and Walter Scott. It was one of the most scenic areas in all of England, with its incomparable lakes, Cumbrian Mountains, and nearby coastal villages.

Now, at Sellafield—entirely owned and operated by

the British Government — P=L=U=T=O=N=I=U=M and Uranium wastes from Europe, Japan and from everywhere, are smelted down in order to extract and reprocess valuable P=L=U=T=O=N-=I=U=M, which is then re-sold to any government that can pay for it. P=L=U=T=O=N=I=U=M has now become the *Gold* that supports what is left of the British Empire, and *is now its chief export to the world.*

The radioactive pollution Sellafield generates is immeasurable. The plant has dumped millions of pounds of hot processing debris into nearby shallow earth trenches and the Irish Sea. It regularly vents P=L=U=T=O=N=I=U=M into the air through its smokestacks and silos. Poisonous concentrations 27,000 times background level, have been measured in the area. P=L=U=T=O=N=I=U=M processing wastes are now regularly dumped into the Atlantic Ocean and the North Sea.

Sellafield is administered by "British Nuclear Fuels," which is an agency of the British Government, and is carefully policed, protected and shielded by "Crown Immunity,"—which means, no citizen or private entity can sue the Government for redress of harm to either life or property. British justice will entertain no complaint against Sellafield. It is further protected under the aegis of "National Defense," and even more importantly, it is entirely shielded from surveillance or oversight by the "Official Secrets Act."

Problems in the area are staggering. Household dust from radiation is 1000 times background levels; one child in 60 on the English and Irish coasts nearby, dies of Leukemia each year; Britain leads the world in lung cancer deaths; radioactive Iodine from plant emissions contaminates the sea, and the local meat, fish and milk; the plant has been dumping over a million gallons of radioactive liquid waste a day into the sea, via its own pipeline to coastal waters.

Sellafield is supervised exclusively by *politically* appointed ministers responsible only to "The Crown." Ireland and Denmark, who have no nuclear power plants, have protested vehemently about the growing air and sea pollution coming their way, but to no avail. Other European nations remain silent, because they are only too eager to have their own nuclear wastes shipped out to England, to come back as usable fuel for their reactors.

British nuclear waste clients constitute a *Who's Who* of technologically advanced countries: Japan, Italy, Germany, Switzerland, Spain, Holland and Sweden. (France has its own waste pipeline into the English Channel at Cap La Hague.)

The nuclear weapons manufacturers of both Britain and the United States have cooperated in building the joint underground nuclear weapons testing facility at Nevada. Since 1962 some 725 underground nuclear explosions have taken place. A great number of them were British. The British Government feels that since there is limited open space in Britain for testing, and, in their view, plenty of it, in the United States, it is better for them to pollute the States, rather than their own country. And, the United States Government agrees with them in this, and has welcomed them here.

We, as the colonial rebel offshoot of "The Mother Country," have yet to dispense with our romantic illusions, our awe, our veneration for Europeans, especially the British, as coming from a gentler, wiser, worldlier, less "materialistic" civilization. The American Academic community in particular, exhibits a singularly uncritical loyalty to Britain. The British nuclear community is almost "invisible," disguised as it is by its seeming reserve and polite gentlemanly manners so admired by Americans who respect public figures with a "low profile."

Because of the success of the British Official Secrets Act, which discourages freedom of information, the dark cloud of secrecy and ignorance has been successful in keeping the people of Britain in a naively trusting and respectful condition about their government's nuclear policies. Indeed, since information on P=L=U=T=O=N=I=U=M is not easily available in England, the people of Cumbria have been obliged to purchase their own Geiger counters to determine the level of radioactive exposure from Sellafield that they and their children are exposed to on a daily basis.

An American writer, Marilynne Robinson, has documented information on Sellafield which she was able to piece together from scattered newspaper reports. Her citations concerning the activities at Sellafield came from *The London*

Times, The Guardian, New Statesman, New Scientist, The Observer. The articles cover the period from 1976 to 1986, and are newspaper accounts of isolated nuclear events, all related to the secret activities at Sellafield. The information in the articles was neither confirmed nor denied by the Government. All that she was able to gather was published in her book *Mother Country* (New York, 1989), to which we refer the interested reader.

DEFENDING THE ENVIRONMENT

What we can do to change government policy on Nuclear Defense, Nuclear Development and the continuing proliferation of P=L=U=T=O=N=I=U=M Contaminated Radioactive Wastes.

We can write to our own Congressional State Senators and Representatives.

The addresses are:

(Your State Senator)
United States Senate
Washington, DC 20515

(Your Congressional Representative)
United States House of Representatives
Washington, DC 20515

WHAT TO ASK FOR:

—Independent Regulatory Oversight of all D.O.E. production facilities and laboratories.

—D.O.E. contractors, and Nuclear Utility Corporations be made fully liable for injury to persons, and for damages to our homes and property.

—Most fundamental of all: United States, as the world's leading superpower, must pursuade all the nations of the world to sign on to a verifiable Nuclear Non-Proliferation Agreement. What harms one nation, vis-à-vis Nuclear, harms us all.

BUT DON'T LEAVE IT TO THE GOVERNMENT:

Governments can no longer initiate what is necessary to our survival. They can only be reactive. Therefore, citizens acting together in groups can accomplish things. Support and join your citizen action committees as the only way to protect your own interests.

NAMES AND ADDRESSES

CARD: Citizens for Alernatives to Radioactive Dumping
144 Harvard SE
Albuquerque, NM 87106

Center for Policy Alternatives
2000 Florida Ave., NW, Suite 400
Washington, DC 20009

Citizen Alert
PO Box 5391
Reno, NV 89513

Concerned Citizens for Nuclear Safety
412 West San Francisco St.
Santa Fe, NM 87501

Conservation International
1015 18th St., NW
Washington, DC 20036

Delaware Valley Greens
740 Catharine St.
Philadelphia, PA 19147

Earth First!
PO Box 5871
Tucson, AZ 85703

Earth Island Institute
300 Broadway, Ste. 28
San Francisco, CA 94133

East Bay Green Alliance
PO Box 3727
Oakland, CA 94609

Environmental Defense Fund
257 Park Ave., S.
New York, NY 10011

Environmental Policy Institute
218 D St., SE
Washington, DC 20003

Green Alliance
661 Aldine
Chicago, IL 60657

Green Committee of Correspondence
PO Box 30208
Kansas City, MO 64112

Greenpeace USA
1436 U St., NW
Washington, DC 20009

International Union for Conservation of Nature and Natural Resources
Gland 1196, SWITZERLAND

The Izaak Walton League of America
1401 Wilson Blvd., Level B
Arlington, VA 22209

National Audubon Society
950 Third Ave.
New York, NY 10022

National Parks and Conservation Association
1015 31st St., NW
Washington, DC 20007

Natural Resources Defense Council
40 West 20th St.
New York, NY 10011

National Wildlife Federation
1400 16th St., SE
Washington , DC 20003

The Nature Conservancy
1815 North Lynn St.
Arlington, VA 22209

New Options, Inc.
PO Box 19324
Washington, DC 20036

Public Citizen
215 Pennsylvania Ave., SE
Washington, DC 20003

Radioactive Waste Campaign
625 Broadway, 2nd Floor
New York, NY 10012

Sierra Club
730 Polk St.
San Francisco, CA 94109

Southwest Research and Information Center
PO Box 4524
Albuquerque, NM 87106

The Wilderness Society
1400 I St., NW 10th Floor
Washington, DC 20005

Worldwatch Institute
1776 Massachusetts Ave., NW
Washington, DC 20036-1904

BIBLIOGRAPHY

Abbey, Edward. *Desert Solitaire.* New York: Simon and Schuster, 1968.

Blumberg, Stanley A. and Gwinn Owens. *Energy and Conflict.* New York: G.P. Putnam's Sons, 1976.

Carson, Rachel. *Silent Spring.* New York: Houghton Mifflin Co., 1962.

Curtis, Richard and Elizabeth Hogan. *Nuclear Lessons.* Harrisburg, Pa.: Stackpole Books, 1980.

Dean, Gordon. *Report on the Atom.* New York: Knopf, 1953.

Epstein, William. *The Last Chance.* New York: The Free Press, 1976.

Falk, Jim. *Global Fission.* Oxford: Oxford Univ. Press, 1983.

Fermi, Laura. *Atoms in the Family: My Life With Enrico Fermi.* Albuquerque: Univ. of New Mexico Press, 1982.

Ford, Daniel F. *Three Mile Island: Thirty Minutes to Meltdown.* New York: Penguin Books, 1982.

Goldschmidt, Bertrand. *The Atomic Complex.* La Grange Park, Ill.: American Nuclear Society, 1980.

Gowing, Margaret. *Britain and Atomic Energy.* London: Macmillan, 1964.

Hilgartner, Stephen, Richard C. Bell and Rory O'Connor. *Nukespeak: Nuclear Language, Visions, and Mindset.* San Francisco: Sierra Club Books, 1982.

Irving, David. *The German Atomic Bomb.* New York: Simon and Schuster, 1967.

Junetka, James W. *City of Fire: Los Alamos and the Atomic Age, 1943–1945.* Albuquerque: Univ. of New Mexico Press, 1979.

Keck, Otto. *Policymaking in a Nuclear Program.* Lexington, Mass.: Lexington, 1981.

Kessler, Edwin. *The Thunderstorm in Human Affairs.* Norman: Univ. of Oklahoma Press, 1981.

Knebel, Fletcher and Charles W. Bailey. *No High Ground.* New York: Bantam Books, 1960.

Lapp, Ralph. *Kill and Overkill.* New York: Basic Books, 1962.

Lillenthal, David. *The Atomic Energy Years.* New York: Harper and Row, 1964.

Lovins, L.H. and A.B. Lovins. *Energy and War.* San Francisco: Friends of the Earth, 1980.

Medvedev, Zhores. *Nuclear Disaster in the Urals.* New York: W.W. Norton, 1979.

McPhee, John. *The Curve of Binding Energey.* New York: Farrar, Strauss & Giroux, 1973.

Miller, Richard L. *Under the Cloud: The Decades of Nuclear Testing.* New York: The Free Press, 1986.

Mojtabai, A.G. *Blessed Assurance: At Home With the Bomb in Amarillo, Texas.* Albuquerque: Univ. of New Mexico Press, 1986.

Moss, Norman. *Men Who Play God.* New York: Harper and Row, 1968.

------------------ *The Politics of Uranium.* London: Andre Deutsch, 1981.

Patterson, Walter C. *The Plutonium Business and the Spread of the Bomb.* San Francisco: Sierra Club Books, 1984.

Pringle, Peter and James Spigelman. *The Nuclear Barons.* New York: Holt, Rinehart and Winston, 1981.

Prins, Gwyn. *Defended to Death.* London: Penguin Books, 1983.

Rappoport, Roger. *The Great American Bomb Machine.* New York: Ballantine, 1971.

Riordan, Michael. *The Day After Midnight: The Effects of Nuclear War.* Palo Alto, Cal.: Cheshire Books, 1982.

Rochlin, Gene I. *Plutonium, Power and Politics.* Berkeley: Univ. of Cal. Press, 1979.

Rosenberg, Howard L. *Atomic Soldiers.* New York: Harper and Row, 1982.

Strauss, Lewis L. *Men and Decisions.* Garden City, N.Y.: Doubleday, 1962.

Szasz, Ferenc Morton. *The Day the Sun Rose Twice.* Albuquerque: Univ. of New Mexico Press, 1984.

Thompson, Theos, and J.G. Beckerley. *The Technology of Nuclear Reactor Safety.* Cambridge, Mass.: MIT Press, 1973.

Uhl, Michael and Tod Ensign. *G.I. Guinea Pigs.* New York: Putnam, 1983.

Walker, William and Mans Lonnroth. *Nuclear Power Struggles.* London: Allen and Unwin, 1983.

Wasserman, Harvey and Norman Soloman. *Killing Our Own: The Disaster of America's Experience With Atomic Radiation.* New York: Delacorte Press, 1982.

Williams, Roger.*The Nuclear Decisions.* London: Croom Helm, 1980.

Willrich, Mason. *International Safeguards and the Nuclear Industry.* Baltimore: Johns Hopkins Univ. Press, 1973.

Wyden, Peter. *Day One: Before Hiroshima and After.* New York: Simon and Schuster, 1984.

STANLEY BERNE is the author of nine published books. In 1982 his novel The Great American Empire *was published in New York to wide critical acclaim. Critic Welch Everman wrote; "Stanley Berne has been a vital force on the American literary scene for three decades, and as a fiction writer, essayist, lecturer and teacher, he has helped to guide the American novel into the late twentieth century."*

Berne's undergraduate degree in psychology is from Rutgers University, and his graduate degree in psychology is from New York University. He was a teaching fellow at Louisiana State University, where he studied for the doctorate in Literature. He has been a research professor in English at Eastern New Mexico University.

The author is widely published in literary journals in the U. S. and abroad, and has been a guest on over 100 TV and radio shows.

He co-hosted and co-produced the nine-part PBS TV series "Future Writing Today." His latest appearances were on "The American Forum Writers' Series" on the "Voice of America" (Washington, D.C.), and on the TV series "Speaking of Language," produced for the Maryland Center for Public Broadcasting.

He lives in a solar home near Santa Fe, New Mexico.

ARLENE ZEKOWSKI'S *guest appearances on 100 TV and radio shows include the "American Forum Writers' Series" on the "Voice of America" and the TV series "Speaking of Language," produced by the Maryland Center for Public Broadcasting.*

"Future Writing Today," the nine part poets and writers interview series, co-hosted and produced by Zekowski and Berne, aired on PBS TV stations across America.

The author of ten published books of fiction, poetry, drama and criticism, her latest book Histories and Dynasties *(Horizon Press, New York) was characterized by* The Times Literary Supplement *(London) as "The newest and most creative writing in English."*

Widely anthologized, Zekowski's latest appearance is in American Writing Today *in which America's prominent authors write about their own work.*

Her latest book, Once A Time Upon P=L=U=T=O=N=I=U=M, *a true fiction-fable about the "nuclear brotherhood," is soon to be published.*